SRINIVASA RAMANUJAN, F.R.S.

A SUPER GENIUS INDIAN MATHEMATICIAN OF 20TH CENTURY, WHO DIED AT THE AGE OF 32

Prof. (Dr.) Jai Paul Dudeja

notionpress
.com

INDIA · SINGAPORE · MALAYSIA

ISBN
Paperback 979-8-90362-567-3
Hardcase 979-8-90362-568-0

"An equation for me has no meaning, unless it expresses a thought of God".

– Srinivasa Ramanujan

Contents

Preface

Dear Readers,

I am extremely happy to see in your hands, this book titled, **"Srinivasa Ramanujan, F.R.S: A Super Genius Indian Mathematician of 20th Century, Who died at the Age of 32"**. It is my firm belief that you have chosen to read this book with a specific aim in mind, and I assure you that you will not be disappointed with it.

Srinivasa Ramanujan (1887–1920) was a self-taught Indian genius who made exceptional contributions to mathematics. Born in Erode, India, he showed awe-inspiring talent as a child prodigy, who mastered advanced trigonometry by the age of 12. In 1914, Ramanujan moved to England to work with G.H. Hardy and was subsequently elected a Fellow of the Royal Society in 1918. Due to poor health, he returned to India and passed away in April 1920 at the age of 32. Ramanujan made pioneering discoveries regarding the properties of partition function, highly efficient series for calculating pi, exponential functions, mock-theta functions, continued fractions, elliptic functions, etc. Ramanujan's work has had a lasting impact, with many of his formulas finding modern applications in fields like cryptography, cancer treatment string theory, black holes, signal processing, polymer chemistry, and C++, etc. His notebooks, filled with thousands of theorems without proofs, have been studied for decades, with many of these proved correct long after his death.

These aspects of Ramanujan are described in this book, which consists of 3 Sections and 5 chapters.

The author would open heartedly love to receive any encouraging/critical comments as well as feedback from the dear readers at his Email ID: drjpdudeja@gmail.com

Sincerely

2026

Prof. (Dr.) Jai Paul Dudeja

Acknowledgements

The seeds of my interest in the continual quest for knowledge were lovingly sown by my revered parents: **Late (Dr.) Shanti Sawrup Dudeja and (Late) Mrs. Jai Devi Dudeja**. I am sure that they are watching me every moment from wherever they are in the other world and continuously sending their blessings to me.

I have greatly benefitted in going through the books, reports, research papers and articles, referred to in the **'Bibliography'** in this Book. I gratefully acknowledge these authors for enhancing my understanding on the subject matter of this book.

Further, I wish to place on record my sincere thanks to the publishing team of the **Notion Press** for their untiring and sincere efforts in bringing out this book in a pleasantly readable form.

Last but not the least, my greatest admiration is reserved for **Mrs. Rita Dudeja**, my wife and my best friend. She is the source of inspiration for me and a co-traveller on the path of trust and truth.

2026

Prof. (Dr.) Jai Paul Dudeja

Profile of the Author

Born in June 1948, Prof. (Dr.) Jai Paul Dudeja holds a brilliant academic record. He did his Master's degree in Physics from Birla Institute of Technology and Science (BITS), Pilani (India), and Ph.D. degree in Physics from the Indian Institute of Technology (IIT), Delhi. He has been in regular employment as a Scientist, Professor, Dean, Director, Principal and a Senior Academic Administrator in various educational institutions, universities, laboratories, public and private organizations. He superannuated as a Senior Scientist and Additional Director in May 2008, in the field of Laser Science & Technology, from the Defence Research and Development Organisation (DRDO), Government of India. After DRDO, he served for over 11 years, till his last posting as a Director, Amity Institute of Laser Technology & Opto-electronics, at Amity University Gurgaon, from where he retired in Nov 2019.

Till date, Dr. Dudeja has published/presented about 90 research papers in various national and international journals/conferences. Besides this, he has authored the following **FORTY Books** on science, technology, spirituality, mysticism, consciousness, and various other subjects:

1. Gayatri Mantra: A GPS to Enlightenment
2. Maha Mrityunjaya Mantra: An Invincible Armour for Conquering Death
3. Ajapa-Japa Sohum-Humsa Mantra: An Eternal Mantra for Inner Consciousness
4. The Third Eye: A Spiritual Laser for Stimulating Inner Awakening
5. Quantum Physics of Consciousness and Non-Duality in Eastern Philosophy
6. Quantum Science of Love, Healing, Happiness, and Bliss in Ancient Wisdom
7. Chakras Healing and Kundalini Awakening by Yogic Techniques,
8. Meditation Practices across the Globe and their Beneficial Effects.
9. Comparative Analysis of Hindu, Buddhist, and Jain Philosophies
10. Mantras for Happiness
11. Om Namah Shivaya: A Powerful Mantra for Mastering Five Elements
12. Trataka: A Concentrated Gazing Technique for Mystical Powers
13. Walking Meditation: Techniques and Benefits
14. Quantum Science of Ganesha Consciousness
15. Tantra Science
16. Quantum Brain, Mind, and Thinking
17. Profound Meditation Techniques in Tibetan Buddhism
18. REIKI: A Holistic Energy Healing Technique

19. Shaktipāt: Instant Transmission of Spiritual Energy from a Siddha Guru to the Disciple
20. Vāstu Shāstra: Ancient Indian Science of Architecture
21. Vedantic Thoughts on Māyā, Mithyā, and the Brahman
22. Siddhis (Supernatural Powers): A Guide for Understanding and Attaining These
23. Near-Death Experience: Scientific Interpretation (Inspired by True Story of a Friend)
24. Universe before the Big Bang: A Deeper Insight into Cosmology
25. Spiritual Intelligence: Significance, Applications, Measurement, and Development

    Techniques
26. Mind-Reading and Artificial Intelligence: Past, Present and Future (Science, Technology,

    Opportunity, Risks and Regulations)
27. Climate Change: A Global Challenge (Causes, Adverse Effects, Monitoring, Adaptation,

    And Mitigation).
28. Converging, Emerging, Innovative, Disruptive, and Critical Technologies for Modern and Future Warfare
29. A Guidebook on Healthy and Unhealthy Foods and Diets: With Description of Over 80 Foods, Diets and Cuisines in the World.
30. Maa Saraswati: The Goddess of Knowledge, Wisdom, Creativity, Speech, Arts, and Music
31. Chanakya: The Great Indian Visionary, Economist, and Strategist: His Legacy and Relevance in Modern Times
32. Navigating Successfully through Maliciously Envious People

33. Role of Sound and Light in Healing, Therapy, and Meditation: A Scientific Analysis
34. How to Quiet Your Mind and Lead a Purposeful Life
35. Vidura and Vidura Niti: Lessons from a Statesman of Mahabharata Times
36. Yoga Nidra: A Guided Relaxation Technique to Induce Yogic Sleep Without Losing Inner Awareness
37. The Anugita: Second Discourse by Lord Krishna to Arjuna; This Time after the End of Kurukshetra War
38. Shiva Sutras of Vasugupta: English Translation and Philosophical Interpretation of God-Consciousness Attainment in Kashmir Shaivism
39. Brahma Sutras: Vedanta Philosophy and the Harmonisation of Upanishads; Verse-by-Verse English Translation, Interpretation and Commentary
40. Aryabhata-I and Aryabhatiya: A Highly Revered Indian Mathematician and Astronomer of the Fifth century, and his Famous Contribution

Dr. Dudeja has delivered many invited lectures on various subjects in the international conferences in India and other countries.

He has been recognized as the “World’s Who’s Who” in Science & Engineering.

SECTION-1

OVERVIEW OF SRINIVASA RANANUJAN (1887-1920)

CHAPTER 1

Life Story of Srinivasa Ramanujan (1887-1913)

1.1 INTRODUCING SRINIVASA RAMANUJAN, A GOD-GIFTED, CHILD PRODIGY,

Super Genius Indian Mathematician of Exceptional Intuition

Srinivasa Ramanujan is one of the greatest mathematicians in the 20th century. In fact, it can be safely said that he is one of the greatest mathematicians in the history. What makes Ramanujan unique is the spectacular and surprising aspect of almost every one of his discoveries and the manner in which he arrived at his results.

Srinivasa Ramanujan Iyengar, F.R.S. (Fellow of Royal Society) in Mathematics, (Born: Erode, near Tanjore), 22 December 1887, at the residence of his maternal grandparents. – Died: Chetput, (Madras), 26 April 1920, at the young age of 32, due to illness) was an exceptionally genius Indian mathematician who worked during the early 20th century. Srinivasa Ramanujan was a mathematician so great that his name transcends jealousies. He was the one superlatively great mathematician. whom India has produced in the last thousand years. His leaps of intuition confound mathematicians even today, several decades after his death. His papers are still plumbed for their secrets. His theorems are being applied in areas of **polymer chemistry, computers, even cancer**-scarcely imaginable during his lifetime. And always the nagging question is: What might have been, had he been discovered a few years earlier, or lived a few years longer?

In September 1887, two months before her child (Ramanujan) was due to be born, a nineteen-year-old Kumbakonam girl named Komalatammal (Ramanujan's mother) travelled to Erode, her parental home, 150 miles upriver, to prepare for the birth of the child she carried. That a woman returned to her native home for the birth of her first child was a tradition widely observed for monitoring vitals by the would-be mother's mother and taking care for the same.

It was here that a son was born to her and her husband Srinivasa, just after sunset on the ninth day of the Indian month of Margasirsha- or Thursday, December 22, 1887. On his eleventh day of life, again in accordance with tradition, the child was formally name, Ramanujan, and almost a year after day after his birth, Srinivasa Ramanujan Iyengar and his mother returned to Kumbakonam, where he would spend most of the next twenty years of his life.

Ramanujan was a simple man. His needs were simple. So were his manners, his humour. He was intelligent in realms outside mathematics, persistent, hardworking, and even, in his own way, charming. But by the lights of Cambridge or, for that matter, of Calcutta or Bombay, he was supremely narrow and naive. He was shamed by the most insignificant slight. His letters, outside their mathematical content, are barren of grace or subtlety.

Ramanujan was a self-taught mathematical prodigy from a town outside Madras, in South India, Despite having almost no formal training in pure mathematics, Ramanujan made substantial contributions to mathematical analysis, number theory, infinite series, and continued fractions, including solutions to mathematical problems then considered unsolvable.

Ramanujan's story is a story of the clash of cultures between India and the West, between the world of Sarangapani Sannidhi Street in Kumbakonam in South India, where Ramanujan grew up, and the glittering world of Cambridge; between the pristine proofs of the

Western mathematical tradition and the mysterious powers of intuition with which Ramanujan dazzled East and West alike.

It is a story of one man and his stubborn faith in his own abilities. In a way, then, this is also a story about social and educational systems, and about how they matter, and how they can sometimes nurture talent and sometimes crush it. How many Ramanujans, dwell in India even today, unknown and unrecognized? And how many in America, Britain and other advanced countries, are locked away in racial or economic ghettos, scarcely aware of worlds outside their own?

This is a story, too, about what you do with genius once you find him. Ramanujan was brought to Cambridge by an English mathematician of aristocratic appearance and high academic credentials, Prof. G. H. Hardy, to whom he had written for help in 1913. Hardy saw that Ramanujan was a rare flower, one not apt to tolerate being stuffed methodically full of all the mathematical knowledge he'd never acquired in India. "I was afraid," Hardy wrote, "that if I insisted unduly on matters which Ramanujan found irksome, I might destroy his confidence and break the spell of his inspiration." Hardy, it turned out, was the third English mathematician to whom Ramanujan had appealed; the other two declined to help. And Hardy did not just recognize Ramanujan's gifts; he went to great lengths to bring him to England, formally train him in the mathematics he had missed, and bring him to the attention of the world.

It was Hardy's ability to see genius in the tattered garb in which Ramanujan was clothed, and to agree that the world was enriched as a result of this. But Hardy was enriched, too. His whole life was shaped by his time with Ramanujan, which he called "the One romantic incident in my life." The story of Ramanujan, then, is a story about two men, and what they meant to each other.

Professor G.H. Hardy, who was Srinivasa Ramanujan's mentor, once remarked that the real tragedy with Ramanujan was not his early death, but that during his most formative years in India prior to his departure to England, his genius was misdirected owing to a lack of formal training. Hardy pointed out that the best original work in any field is usually done before the age of 30 and that this is especially true of mathematics. So, if Ramanujan had lived longer, although he would have produced a good deal more, Hardy argued that the quality of his work would not have been greater. Several years after Ramanujan's death we have come to realise that Hardy may have well been wrong in this estimation of Ramanujan, for the depth and power of Ramanujan's mathematics actually increased with the years. His work in England after contact with Hardy and exposure to more sophisticated techniques shows a greater amount of variety, advanced level and power.

Ramanujan was a man who grew up praying to stone deities; who for most of his life took counsel from a family goddess, the **goddess Namagiri of Namakkal**, declaring it was she to whom his mathematical insights were owed; whose theorems would, at intellectually backbreaking cost, be proved true--yet leave mathematicians baffled that anyone could divine them in the first place. This is also story about an uncommon and individual mind, and what may suggest about creativity, intuition, and intelligence, combined in one.

Ramanujan initially developed his own mathematical research in isolation. He tried to attract the attention and interest of the leading professional mathematicians of the world in his work, but failed for the most part. What he had to show them was too novel, too unfamiliar, and presented in unusual ways; they could not be bothered." Seeking mathematicians who could better understand his work, in 1913, he sent letters to the English mathematician Prof. G. H. Hardy at the University of Cambridge.

1.2 BIRTH AND EARLY LIFE OF RAMANUJAN

The **Cauvery river** was a familiar, recurring constant of Ramanujan's life. At some places along its length, palm trees, their trunks heavy with fruit, leaned over the river at different angles. At others, leafy trees formed a canopy of green over it, the knotted roots snaking along the riverbank. During the monsoon, its waters might rise ten, fifteen, or twenty feet high, sometimes drowning cattle allowed to graze too long beside it. Come the dry season, the torrent became a memory, the riverbanks' wide sandy beaches, and the Cauvery river itself but a feeble trickle tracing the deepest channels of the riverbed.

Kumbakonam, Ramanujan's hometown, flanked by the Cauvery river and one of its tributaries, lay in the heartland of staunchly traditional South India, 160 miles south of Madras, in the district then known as Tanjore. Half the district's thirty-seven hundred square miles, was watered directly by the river.

Dakshin Gange, the Cauvery river was called-the Ganges of the South. And it made the delta the most densely populated and richest region in all of South India. The whole edifice of the region's life, its wealth as well as the rich spiritual and intellectual lives, its wealth encouraged, all depended on its waters. The Cauvery river was a place for spiritual cleansing; for agricultural surfeit; for drawing water and bathing each morning; for cattle, led into its shallow waters by men in white dhotis and turbans, to drink; and always, for women, standing knee-deep in its waters, to let their snaking ribbons of cotton or silk drift out behind them into the gentle current, then gather them up into sodden clumps of cloth and slap them slowly, relentlessly, against the water-worn rocks.

Ramanujan (literally meaning, "younger brother of Rama", a Hindu deity) was born in the British-ruled India on 22 December 1887 into a Tamil Brahmin Iyengar family in **Erode**, in modern Tamil Nadu. His father, Kuppuswamy Srinivasa Iyengar, originally from Thanjavur district, worked as a clerk in a **sari shop**. His mother, **Komalatammal**, was a housewife and sang at a local temple. Ramanujan's mother, Komalatammal, sang bhajans, or devotional songs, at a nearby temple. Half the proceeds from her group's performances went to the temple, the other half to the singers.

With her husband earning only about twenty rupees per month, the five or ten rupees, she earned this way mattered; never would she miss a rehearsal. They lived in a small traditional home on Sarangapani **Sannidhi Street** in the town of **Kumbakonam**. The family home became a museum subsequently. When Ramanujan was a year and a half old, his mother gave birth to a son, **Sadagopan, who died less than three months later.**

In December 1889, Ramanujan's mother was missing from her singing group for some days , four or five in a row. So, one day, the head of the singing group showed up at Komalatammal's house to investigate. She saw that two-year-old Ramanujan was suffering from smallpox.

Ramanujan would bear the scars of his childhood smallpox all his life. But he recovered, and in that was fortunate. For in Tanjore District, around the time he was growing up, a bad year for smallpox meant four thousand deaths. Fewer than one person in five was vaccinated. A cholera epidemic when Ramanujan was ten, the cholera killed fifteen thousand people. Three or four children in every ten died before they'd lived a year.

Ramanujan moved with his mother to her parents' house (Ramanujan's maternal grandfather) in Kanchipuram, near Madras (modern Chennai). When Ramanujan was almost four, in November 1891, a girl was born. By the following February, she, too, was dead. When Ramanujan was six and a half, his mother gave birth to yet another child, Seshan-who also died before the year was out. Much later, **two brothers** did survive-**Lakshmi Narasimhan, born in 1898**, when Ramanujan was ten, and **Tirunarayanan**, born when he was even teen. But the death of his infant brothers and sister during those early years meant that he grew up with the solicitous regard and central position of an only child.

For Ramanujan's first three years, he scarcely spoke. Perhaps, it is tempting to think, because he simply didn't choose to; he was an enormously self-willed child. It was common in those days for a young wife to shuttle back and forth between her husband's house and that of her parents, and Komalatammal, worried by her son's muteness, took Ramanujan to see her father, then living in Kanchipuram, near Madras.

There, at the urging of an elderly friend of her father's, Ramanujan began the ritual practice of *Akshara Abhyasam*: his hand, held and guided by his grandfather, was made to trace out Tamil characters in a thick bed of rice spread across the floor, as each character was spoken aloud.

Soon fears of Ramanujan's dumbness were dispelled and **he began to learn the 12 vowels, 18 consonants, and 216 combined consonant-vowel forms of the Tamil alphabet.**

On October 1, 1892, the traditional opening day of school, known as Vijayadashami, he was enrolled, to the accompaniment of ancient Vedic chants, in the local pial school. A pial is the little porch in front of most South Indian houses; a pial school was just a teacher meeting there with half a dozen or so pupils.

In October 1892, Ramanujan was enrolled at the local school. After his maternal grandfather lost his job as a court official in Kanchipuram, Ramanujan and his mother moved back to Kumbakonam, and he was enrolled in **Kangayan Primary School**. At Kangayan Primary School, Ramanujan performed well.

After the death of his paternal grandfather, who had suffered from leprosy, Ramanujan, seven at that time, broke out in a bad case of itching and boils. But this was not the first hint of a temperament inclined to extreme and unexpected reactions to stress. Indeed, Ramanujan was a sensitive, stubborn, and-if a word more often reserved for adults in their prime can be applied to a little boy-eccentric child. While yet an infant back in Erode, he wouldn't eat except at the temple. He was sent back to his maternal grandparents, then living in Madras. He did not like the school in Madras, and tried to avoid attending. **His family employed a local**

constable to make sure that he attended the school. Within six months, Ramanujan was back in Kumbakonam.

For about two years, Ramanujan was shuffled between schools. Beginning in March 1894, while still at his mother's parents' house in Kanchipuram, he briefly attended a school in which the language of instruction was not his native Tamil but the related but distinct Telugu. There, sometimes punished by having to sit with his arms folded in front of him and one finger turned up to his lips in silence, he would at times stalk out of class in a huff.

By mid-1895, after an unhappy six months' period in Madras, Ramanujan was once more back in Kumbakonam. South India was not always hot; but it was never cold either. At a latitude of about eleven degrees north. Kumbakonam lay comfortably within the tropics; even on a January winter's night, the thermometer dropped, on an average, only to 22 degrees Celsius. And that climatological fact established an architectural fact, for it gave South Indian homes a kind of permeability. Windows were there, but these were merely cutouts in the wall, perhaps with bars or shutters, never space-sealing panes of glass that left you conscious of being on one side or another. In the middle of most houses was a small courtyard, open to the sky-like a skylight but again without the glass-that brought rain into the centre of the house, where it was funnelled to a drain that led back outside. In Ramanujan's house, smells from outside wafted inside. Lizards crawled; and mosquitos flew unimpeded. The soft South Indian air, fragrant with roses, with incense, with cow dung burned as a fuel, wafted over everything. When Ramanujan was six, the town completed a drainage system. But this was designed to carry off only the surface water, not sewage, and most of the town's health problems continued unabated.

Another specialty of the town was the silk saris, the product of two thousand small looms manned by three thousand people. No place in South India was better known for its fine silk saris, dazzling in bright colours, embroidered in silver stripes, fringed with gold, than Kumbakonam and neighbouring Tanjore. Saris woven in Kumbakonam could cost as much as a hundred rupees-a year's income to many poor families. **Ramanujann' father (Srinivasa Iyengar) worked in one of these sari shops as a clerk.** Normally, such a clerk remained one all his life-waiting on customers, taking orders, performing routine paperwork, perhaps travelling to nearby villages to collect money from people who bought sarees from his shop. Occasionally a clerk might be taken into the business or would go off to start his Own. But that required some special drive or entrepreneurial temperament. Apparently Srinivasa was good at appraising fabrics, a skill upon which his employer relied; but beyond that, whatever it took to step to a better job he could never muster.

Day-after-day, year-after-year, Ramanujan's father was at the shop, largely absent from his son's early life. Indian society generally left the father little role to play at home, casting him as an aloof, physically undemonstrative, even unwelcoming figure whose relationship with his children was largely formal. Srinivasa Iyengar was almost invisible, his name largely absent from the family accounts. "Very quiet," a boyhood friend of Ramanujan recalled him. Someone else would resort to the word "weightless." But even had he been otherwise, he could scarcely have competed with Komalatammal (Ramanujan's mother) as an influence on their son.

Since Ramanujan's father was at work most of the day, his mother took care of the boy, and they had a close relationship. From her, he learned about tradition and puranas (ancient Hindu texts), to singing religious songs, attended pujas at the temple, maintained particular eating habits—all part of Brahmin culture. She poured prodigious energy into her spiritual life. In Hindu families, the women were apt to be more pious, and more scrupulous about observing tradition, than the men. So, it had been in her own family; her mother was said to have gone into hypnotic trances that placed her in communion with the gods. And so, Town High School it was in Ramanujan's family. Komalatammal was fiercely devout, held prayer meetings at her home, sangat the temple, pursued astrology and palmistry.

Always, the name of their family deity, the Goddess Namagiri of Namakkal, was on her lips. "An exceptionally gifted lady with psychic powers and a remarkable imagination" was how one friend of the family described her. She had "'a remarkable repertoire of mythological stories and used to tell me stories from [the] ancient Mahabharata and Ramayana to [the] later Vikramaditya legends." Any pause in the telling these stories, was cause for yet another murmured appeal to the goddess Namagiri.

Quiet and contemplative, Ramanujan was fond of asking questions like, 'Who was the first man in the world?' Or, 'How far is it between clouds?' He liked to be by himself, a tendency abetted by parents who, when friends called, discouraged him from going out to play; so, he'd talk to them from the window overlooking the street. He lacked all interest in sports. And in a world where obesity was virtually unknown, where bones protruded from humans and animals alike, he was, first as a child and then for most of his life, little bit fat. He used to say-whether as boast, joke, or lament remains unclear-that if he got into a fight with another boy he had only to fall on him to crush him to pieces.

Just before turning 10, **in November 1897, he passed his primary examinations in English, Tamil, geography, and arithmetic with the best scores in the district. That year, Ramanujan entered the Town Higher Secondary School,** where he encountered formal mathematics for the first time.

A child prodigy by the age of 11, Ramanujan had exhausted the mathematical knowledge of two college students who were the lodgers at his home. He was later lent a book written by **S. L. Loney** on advanced trigonometry. He mastered this by the age of 12 while discovering sophisticated theorems on his own. By the age of 14, he received merit certificates and academic awards that continued throughout his school career, and he assisted the school in the logistics of assigning its 1,200 students (each with differing needs) to its approximately 35 teachers. **He completed mathematical exams in half the allotted time,** and showed a familiarity with geometry and infinite series. Ramanujan was shown how to solve cubic in 1902. He would later develop his own method to solve the quartic equations (polynomial equations with degree 4). In 1903, he tried to solve the quintic equations (polynomial equations with degree 5), not knowing that it was impossible to solve with radicals.

In 1903, when he was 16, Ramanujan obtained from a friend a library copy of **'A Synopsis of Elementary Results in Pure and Applied Mathematics': G. S. Carr's collection of over 5,000 theorems.** Ramanujan studied the contents of the book in detail. The next year, Ramanujan independently developed and investigated the **Bernoulli numbers** and calculated the **Euler–** By 1904 Ramanujan had begun to undertake deep research. He investigated the series $\sum\left(\frac{1}{n}\right)$ and calculated Euler's constant to **15 decimal places**. He began to study the **Bernoulli numbers**, although this was entirely his own independent discovery.

When Ramanujan graduated from the Town Higher Secondary School in 1904, he was awarded the K. Ranganatha Rao prize for mathematics by the school's headmaster, Krishnaswami Iyer. Mr. Iyer introduced Ramanujan as an outstanding student **who deserved scores higher than the maximum.** Ramanujan received a scholarship to study at Government Arts College, Kumbakonam, but was so intent on mathematics that he could not focus on any other subjects and failed in most of them, **losing his scholarship in the process**. In August 1905, **Ramanujan ran away from home**, heading towards **Visakhapatnam**, and stayed in **Rajahmundry** for about a month. He later enrolled at **Pachaiyappa's College in Madras**. There he passed in mathematics, choosing only to attempt questions that appealed to him and leaving the rest unanswered, but performed poorly in other subjects, such as English, physiology, and Sanskrit. Ramanujan failed his F.A. exam in December 1906 and again a year later. Without an F,A. degree, he left college and continued to pursue independent research in mathematics, living in extreme poverty and often on the brink of starvation.

1.3 ADULTHOOD OF RAMANUJAN IN INDIA

On 14 July 1909, Ramanujan **married Janaki** (Janakiammal; born: 21 March 1899 – died:13 April 1994), a girl which Ramanujan's mother had selected for him a year earlier and who was ten years old when they married. It was not unusual then for marriages to be arranged with girls at a young age. Janaki was from Rajendram, a village close to Marudur (Karur district) Railway Station. Ramanujan's father did not participate in the marriage ceremony (Was he too busy in his saree shop or some other reason?). As was common at that time, Janaki continued to stay at her maternal home for three years after marriage until she reached puberty. In 1912, Janaki and Ramanujan's mother joined Ramanujan in Madras.

In 1910, after a meeting between the 23-year-old Ramanujan and the founder of the Indian Mathematical Society, V. Ramaswamy Aiyer, Ramanujan began to get recognition in Madras's mathematical circles, leading to his inclusion as a researcher at the University of Madras.

After the marriage, Ramanujan developed a **hydrocele testis**. The condition could be treated with a routine surgical operation that would release the blocked fluid in the scrotal sac, but his family could not afford the operation. **In January 1910, a doctor volunteered to do the surgery at no cost.**

After his successful surgery, Ramanujan searched for a job. He stayed at a friend's house while he went from door to door around Madras looking for a clerical position. To make money, he tutored students at the Presidency College who were preparing for their Fellow of Arts (F.A.) exam.

In late 1910, Ramanujan was sick again. He feared for his health, and told his friend R. Radhakrishna Iyer to "hand his notebooks over to Professor Singaravelu Mudaliar (the mathematics professor at Pachaiyappa's College) or to the British professor Edward B. Ross, of the Madras Christian College."

After Ramanujan recovered and retrieved his notebooks from Iyer, he boarded a north-bound train from Kumbakonam and, about halfway to Madras, got off at Villupuram, just west of Pondicherry, the coastal city then still in French hands. At Villupuram, he changed trains for the twenty-mile trip west to Tirukoilur, a town of about nine thousand that was headquarters of its district. In Tirukoilur, V. Ramaswami Iyer held the mid-level government post of deputy collector. (Iyer, also spelled as Aiyar, was the caste name of Brahmins who worshipped Shiva, and was ubiquitous in South India.) What made Ramaswami especially worth traveling to see was that he was a mathematician; in particular, he had recently founded the Indian Mathematical Society. Everyone called him "Professor," though he held no academic post. Back while a student at Presidency College, it seems, he had contributed mathematical articles to the Educational Times in England. Its editors, assuming he was a college professor, addressed him as such, and the name stuck.

In May 1913, upon securing a research position at Madras University, Ramanujan moved with his family to Triplicane.

1.4 PURSUIT OF CAREER IN MATHEMATICS AND EMPLOYMENT IN INDIA

In, Ramanujan met deputy collector V. Ramaswamy Aiyer, who founded the Indian Mathematical Society. Wishing for a job at the revenue department where Aiyer worked, Ramanujan showed him his mathematics notebooks. As Aiyer later recalled:

"I was struck by the extraordinary mathematical results contained in (the notebooks of Ramanujan). I had no mind to smother his genius by an appointment in the lowest rungs of the revenue department."

Aiyer sent Ramanujan, with letters of introduction, to his mathematician friends in Madras. Some of them looked at his work and gave him letters of introduction to R. Ramachandra Rao, the district collector for Nellore and the secretary of the Indian Mathematical Society. Rao was impressed by Ramanujan's research but doubted that it was his own work. Ramanujan mentioned a correspondence he had with Professor Saldhana, a notable Bombay mathematician, in which Saldhana expressed a lack of understanding of his work but concluded that he was not a fraud. Ramanujan's friend C. V. Rajagopalachari tried to quell Rao's doubts about Ramanujan's academic integrity. Rao agreed to give him another chance, and listened as Ramanujan discussed **elliptic integrals, hypergeometric series, and his theory of divergent series,** which Rao said ultimately convinced him of Ramanujan's brilliance. When Rao asked him what he wanted, Ramanujan replied that he needed work and financial support. Rao consented and sent him to Madras. He continued his research with Rao's financial aid. With Aiyer's help, Ramanujan had his work published in the *Journal of the Indian Mathematical Society.*

As the Journal editor M. T. Narayana Iyengar noted:

Mr. Ramanujan's methods were so terse and novel and his presentation so lacking in clearness and precision, that the ordinary (mathematical reader), unaccustomed to such intellectual gymnastics, could hardly follow him.

In 1911 Ramanujan approached the founder of the Indian Mathematical Society for advice on a job. After this he was appointed to his first job, a temporary post in the Accountant General's Office in Madras. It was then suggested that he approach Ramachandra Rao who was a Collector at Nellore. Ramachandra Rao was a founder member of the Indian Mathematical Society who had helped start the mathematics library. He writes:

"A short uncouth figure, stout, unshaven, not over clean, with one conspicuous feature – shining eyes – walked in with a frayed notebook under his arm. He was miserably poor. ... He opened his book and began to explain some of his discoveries. I saw quite at once that there was something out of the way; but my knowledge did not permit me to judge whether he talked sense or nonsense. ... I asked him what he wanted. He said he wanted a pittance to live on so that he might pursue his researches."

Ramanujan later wrote another paper and also continued to provide problems in the above Journal. In early 1912, he got a temporary job in the Madras Accountant General's office, **with a monthly salary of 20 rupees.** He carried on only a few weeks. Toward the end of that assignment, he applied for a position under the Chief Accountant of the Madras Port Trust.

In a letter dated 9 February 1912, Ramanujan wrote to the Chief Accountant of the Madras Port Trust :

Sir,

I understand there is a clerkship vacant in your office, and I beg to apply for the same. I have passed the Matriculation Examination and studied up to the F.A. but was prevented from pursuing my studies further owing to several untoward circumstances. I have, however, been devoting all my time to Mathematics and developing the subject. I can say I am quite confident I can do justice to my work if I am appointed to the post. I therefore beg to request that you will be good enough to confer the appointment on me.

Attached to his application was a recommendation from E. W. Middlemast, a mathematics professor at the Presidency College, who wrote that Ramanujan was "a young man of quite exceptional capacity in Mathematics". Three weeks after he applied, on 1 March 1912, Ramanujan learned that he had been accepted as a Class III, Grade IV accounting clerk**, making 30 rupees per month.** At his office, Ramanujan easily and quickly completed the work he was given and spent his spare time doing mathematical research. Ramanujan's boss, Sir Francis Spring, and S. Narayana Iyer, a colleague who was also treasurer of the Indian Mathematical Society, encouraged Ramanujan in his mathematical pursuits.

CHAPTER 2

Efforts to Travel to England for Study (1913-14)

2.1 CONTACTING BRITISH MATHEMATICIANS

In the spring of 1913, Narayana Iyer, Ramachandra Rao and E. W. Middlemast tried to present Ramanujan's work to **British mathematicians. M. J. M. Hill of University College London**, who commented that Ramanujan's papers were riddled with holes. He said that although Ramanujan had "a taste for mathematics, and some ability", he lacked the necessary educational background and foundation to be accepted by mathematicians. Although Hill did not offer to take Ramanujan on as a student, he gave A thorough and serious professional advice on his work. With the help of friends, Ramanujan drafted letters to leading mathematicians at Cambridge University.

The **first two professors, H. F. Baker and E. W. Hobson,** returned Ramanujan's papers without comment.

2.2 RAMANUJAN'S LETTERS TO HARDY

The two letters Ramanujan wrote to Hardy in 1913 are considered to be among the greatest in mathematical history. The letters not only contained a fantastic collection of spectacular formulae, but also profoundly influenced the mathematical careers of both Hardy and Ramanujan. Hardy's initial reaction on seeing the letters was that Ramanujan was a fraud because many of the formulas were known, some were incorrect, and there were no hints of proofs. But then there were several astonishingly beautiful formulas that were correct and very deep. Only a mathematician of the highest class could write them down. So, on second thought Hardy concluded that it was more probable that Ramanujan was a genius and unlikely that he was a fraud because no one but a true genius could have the imagination to invent such formulae. The great philosopher Bertrand Russell says that one evening in Trinity College he found the usually placid Hardy in a wild state of excitement talking about a new Euler or Jacobi from India! Hardy was convinced that Ramanujan was wasting his time in India rediscovering past work and would profit immensely by coming into contact with professional mathematicians. He thus invited Ramanujan to Cambridge University to work with him. Although Ramanujan's mother initially resisted this, she eventually realised that she should not stand in the way of her son's progress and so gave Ramanujan permission to go to England.

On 16 January 1913, Ramanujan wrote to **Prof. G. H. Hardy**, whom he knew from studying *Orders of Infinity* (1910). Coming from an unknown mathematician, the nine pages of mathematics made Hardy initially view Ramanujan's manuscripts as a possible fraud. Hardy recognised some of Ramanujan's formulae but others "seemed scarcely possible to believe".

The first result had already been determined by G. Bauer in 1859. The second was new to Hardy, and was derived from a class of functions called hypergeometric series, which had first been researched by Euler and Gauss. Hardy found these results "much more intriguing" than Gauss's work on integrals. After seeing Ramanujan's theorems on continued fractions on the last page of the manuscripts, Hardy said the theorems "defeated me completely; I had never seen anything in the least like them before", and that they "must be true, because, if they were not true, no one would have the imagination to invent them". **Hardy asked a colleague, J. E. Littlewood,** to take a look at the papers. Littlewood was amazed by Ramanujan's genius. After discussing the papers with Littlewood, Hardy concluded that the letters were "certainly the most remarkable I have received", and that

Ramanujan was "a mathematician of the highest quality, a man of altogether exceptional originality and power". One colleague, **E. H. Neville**, later remarked that "No one who was in the mathematical circles in Cambridge at that time can forget the sensation caused by this letter... not one [theorem] could have been set in the most advanced mathematical examination in the world".

Is sound knowledge of related fields necessary for research? A moot question indeed! A sophisticated mathematician looks at a problem not in isolation, but in terms of its relationship with other questions and then chooses the appropriate tools to solve it. Sometimes when such approaches have not made headway, a radically new idea from a relatively inexperienced researcher produces a breakthrough. **Ramanujan was an extreme example of a mathematician whose contributions were of such high calibre that they belied his lack of formal training.** But the penalty he paid was that much of his work turned out to be rediscovery. Whether Ramanujan would have reached greater heights had he been provided rigorous training is debatable. Hardy was of the opinion that such training would have made Ramanujan less of a genius. Instead of taking sides on this issue, we make note that the brilliance of Ramanujan combined with the sophistication of Hardy was the key to their successful collaboration.

On 8 February 1913, Hardy wrote to Ramanujan a letter expressing interest in his work, adding that it was "essential that I should see proofs of some of your assertions". Before his letter arrived in Madras during the third week of February, Hardy contacted the Indian Office to plan for Ramanujan's trip to Cambridge. Secretary Arthur Davies of the Advisory Committee for Indian Students met with Ramanujan to discuss the overseas trip. **In accordance with his Brahmin upbringing, Ramanujan refused to leave his country to "go to a foreign land", and his parents were also opposed for the same reason.** Meanwhile, he sent Hardy a letter packed with theorems, writing, "I have found a friend in you who views my labour sympathetically."

To supplement Hardy's endorsement, Gilbert Walker, a former mathematical lecturer at Trinity College, Cambridge, looked at Ramanujan's work and expressed amazement, urging the young man to spend time at Cambridge. As a result of Walker's endorsement, B. Hanumantha Rao, a mathematics professor at an engineering college, invited Ramanujan's colleague Narayana Iyer to a meeting of the Board of Studies in Mathematics to discuss "what we can do for S. Ramanujan". The board agreed to grant Ramanujan a **monthly research scholarship of 75 rupees for the next two years** at the University of Madras.

While he was engaged as a research student, Ramanujan continued to submit papers to the *Journal of the Indian Mathematical Society*. In one instance, Iyer submitted some of Ramanujan's theorems on summation of series to the journal, adding, "The following theorem is due to S. Ramanujan, the mathematics student of Madras University." Later, in November, British Professor Edward B. Ross of Madras Christian College, whom Ramanujan had met a few years before, stormed into his class one day with his eyes glowing, asking his students, **"Does Ramanujan know Polish?"** The reason was that in one paper, Ramanujan had anticipated the work of a Polish mathematician whose paper had just arrived in the day's mail. In his quarterly papers, Ramanujan drew up theorems to make definite integrals more easily solvable. Working off Giuliano Frullani's 1821 integral theorem, Ramanujan formulated generalisations that could be made to evaluate formerly unyielding integrals.

Hardy's correspondence with Ramanujan soured after Ramanujan refused to come to England. Hardy enlisted a colleague lecturing in Madras, **E. H. Neville**, to mentor and bring Ramanujan to England. Neville asked Ramanujan why he would not go to Cambridge. Ramanujan apparently then accepted the proposal; Neville said, "Ramanujan needed no converting" and "his parents' opposition had been withdrawn". Apparently, Ramanujan's mother had a vivid dream in which Ramanujan was surrounded by Europeans, and the family goddess, the deity of **Namagiri** (Goddess Mahalakshmi) commanded her "to stand no longer between her son and the fulfilment of his life's purpose".

2.3 DEPARTURE OF RAMANUJAN FOR ENGLAND IN 1914

Ramanujan departed for England from Madras aboard the **ship S.S. Nevasa** on 17 March 1914. leaving his wife to stay with his parents in India. He disembarked in London on 14 April 1914.

The ship had been designed expressly for service in the tropics. Yet Ramanujan could find no comfort aboard her, no relief from the rough seas that left him, on his first ocean voyage, seasick and unable to eat. Temporary relief came with the ship's first stop in Colombo, capital of Ceylon (today's Sri Lanka), the large island just off India's southeast coast. Long before the ship Nevasa docked, passengers could smell Colombo's cinnamon gardens. On 19 March 1914, the ship steamed out of port of Colombo, skirted south of Cape Comorin, at the tip of the subcontinent, and made direct journey for Aden, a week's passage across the Arabian Sea.

For Ramanujan, now past his sea-sickness, the voyage grew pleasant. Now he could enjoy the ship's roominess. He had his vegetarian food. He'd met several among his two hundred or so fellow passengers. Sometimes, the story goes, he withdrew to his second-class cabin and played with numbers suggested by the dimensions of the cabin or the number of passengers.

Ramanujan was not an introspective man, unlike Hardy would put it, "particularly interested in his own history or psychology." But now, in the coat and collar that tortured him, amidst a limitless expanse of sea, as day by day the ship Nevasa steamed west at fourteen knots and the fresh sea breezes swept across the deck.

After a week of steady steaming, more than two thousand miles out of Colombo, the ship Nevasa docked in at Aden, at the southernmost end of the Red Sea. Now began the part of the journey that many among the English passengers, at least, had learned to dread-the fourteen-hundred-mile passage up the Red Sea, where day after day the ship baked in a desert heat that never deviated much from a 32 degrees Celsius. The passage gave the language a new word: if, outward-bound to India, you got stuck in a **starboard** cabin (the side of a ship that is on the right when you are facing towards its front), you absorbed the full heat of the sun all afternoon and by bedtime your cabin smouldered; **portside** cabins (opposite of starboard cabin), meanwhile, had the whole afternoon to cool down. On the homeward passage, it was all reversed. Not surprisingly, VIPs managed to land the cooler cabins, which were designated "**Port Outward-Starboard Homeward" and granted the acronym POSH.**

The next day, the ship Nevasa sailed from Port and steamed into the Mediterranean. On April 7, 1914, after a stop in Genoa, it was through the Strait of Gibraltar and up along the Spanish coast through the Bay of Biscay to England. The ship Nevasa docked first at Plymouth, then steamed up the English Channel and arrived at the mouth of the Thames on April 14, 1914.

Waiting for Ramanujan at the dock was **Neville and his older brother**, who had arrived there by car. They drove to 21 Cromwell Road, in the South Kensington district of London, a reception centre for Indian students just arrived in England.

CHAPTER 3

Life in England for Ramanujan (1914-1919)

3.1 LIFE IN ENGLAND FOR RAMANUJAN

Ramanujan departed from Madras aboard the S.S. Nevasa on 17 March 1914. When he disembarked in London on 14 April. Four days later, Neville took him to his house on Chesterton Road in Cambridge.

Just off the boat and England was already a strange new world for any Indian. Cromwell Road, to which Neville now took Ramanujan, was supposed to ease the transition. Of course it didn't. The National India Association had offices there. Several rooms in the stately Georgian-styled corner buildings were available to students passing through. And across the street, the imposing edifice of the Natural History Museum, fairly glowing in its two-toned stone and adorned with sculpted griffins, lent an appropriately imperial lustre to the immigrant's first days in England. But Cromwell Road typically failed in its mission; a study a few years later would chide the reception centre for invariably making things worse for newcomers, not better. But unlike most Indians in England, Ramanujan had by his side, in Neville, a Cambridge don to help smooth the way. Ramanujan also met A. S. Ramalingam, a twenty-three-year-old engineer from Cuddalore, south of Madras, who had been in England for four years and who also tried to help him feel at home. In any case, Ramanujan survived whatever rigours Cromwell Road could inflict; and on April 18, 1914, he went with Neville to Cambridge. Soon he was settled in Neville's house on

Chestertown Road, in a little suburb of Cambridge just across the River Cam from the town itself.

Ramanujan immediately began his work **with Littlewood and Hardy**. After six weeks, Ramanujan moved out of Neville's house and took up residence on Whewell's Court, a five-minute walk from Hardy's room.

Ramanujan had not come to Cambridge to go to school. But arriving in time for the Easter term, which began in late April (1914), he did attend a few lectures. Some were Hardy's. Others, on elliptic integrals, were given by **Arthur Berry**, a King's College mathematician in his early fifties. One morning early in the term, Berry stood at the blackboard working out some formulas and at one point looked over to Ramanujan, whose face fairly glowed with excitement. Was he, Berry inquired, following the lecture? Ramanujan nodded. Did he care to add anything? At that, Ramanujan stood, went to the blackboard, took the chalk, and wrote down results Berry had not yet proved and which, Berry concluded later, he could not have known before. Soon the word was spreading out about Ramanujan.

W. N. **Bailey,** then an undergraduate, heard "strange rumours that he had been unable to pass examinations in India, and that he had run away from such terrors. But apart from these rumours, “we only knew that his name was Ramanujan, and even this was pronounced wrongly," probably Rah-ma-noo-jn. People didn't often see him; he was usually busy in his rooms. But when they did, they noticed him-remembered his squat, solitary figure as, in the words of one, he "waddled" across Trinity's Great Court, his feet in slippers, unable yet to wear Western shoes.

Hardy and Littlewood began to look at Ramanujan's notebooks. Hardy had already received 120 theorems from Ramanujan in the first two letters, but there were many more results and theorems in the notebooks. Hardy saw that some were wrong, others had already been discovered, and the rest were new breakthroughs.

Ramanujan left a deep impression on Hardy and Littlewood. Littlewood commented, "I can believe that he's at least a Jacobi", while Hardy said he "can **compare Ramanujan only with Euler or Jacobi**." There is much in common regarding the kind of mathematics Euler, Jacobi, and Ramanujan worked on.

Ramanujan saw Hardy nearly every day and could show him the methods he had developed in India that he'd been loath to describe by international post. Meanwhile, Hardy had the notebooks themselves before him and, with their author by his side, could study them as much as he wished.

Many of the 120 theorems Ramanujan had sent Hardy in those first two letters, he (Hardy) could see now, had been plucked intact from the notebooks. Here, in chapter 5, section 30 of the second notebook, was what Ramanujan had written in the first letter about a class of numbers built up from "an odd number of dissimilar prime divisors." From chapter 5 also came much of the work going into Ramanujan's first published paper on **Bernoulli numbers**. In chapter 6 was that bizarre stuff from the first letter about divergent series that, Ramanujan had feared, might persuade Hardy he was destined for the lunatic asylum-the one where I +2+ 3 +4+ ..... unaccountably added up to - **1/12**. On its face, that was ridiculous; yet it sought to give meaning to divergent series-which at first glance added up to nothing more revealing or precise than infinity. But now Hardy found something like Ramanujan's reasoning behind it, which involved a "constant" that, as Ramanujan wrote, "is like the centre of gravity of a body": a concept borrowed from, of all places, elementary physics.

A few of Ramanujan's results were, Hardy could see, wrong. Some were not as profound as Ramanujan would have liked to think. Some were independent rediscoveries of what Western mathematicians had found fifty years before, or a hundred, or two. But many-perhaps a third, Hardy would reckon, perhaps two-thirds, later mathematicians would estimate--were breathtakingly new. Ramanujan's fat, mathematics-rich letters, Hardy now saw, represented but the thinnest sampling, the barest tip of the iceberg, of what had accumulated over the past decade in his notebooks. There were thousands of theorems, corollaries, and examples. Maybe three thousand, maybe four. For page-after-page, they stretched on, rarely watered down by proof or explanation, almost aphoristic in their compression, all their mathematical truths boiled down to a line or two.

Ramanujan spent nearly five years in Cambridge collaborating with Hardy and Littlewood, and published part of his findings there.

Hardy and Ramanujan had highly contrasting personalities. Their collaboration was a clash of different cultures, beliefs, and working styles. In the previous few decades, the foundations of mathematics had come into question and the need for mathematically rigorous proofs was recognised. Hardy was an atheist and an apostle of proof and mathematical rigour, whereas Ramanujan was a deeply religious man who relied very strongly on his intuition and insights. Hardy tried his best to fill the gaps in Ramanujan's education and to mentor him in the need for formal proofs to support his results, without hindering his inspiration—a conflict that neither found easy.

For five years, walled off from India by the war (1st world war), Ramanujan would remain in strange, cold, distant England, fashioning, through twenty-one major papers, an enduring mathematical legacy. Then, he would go home to India (in 1919), to a hero's welcome, and die (on 26 April 1920).

Ramanujan was awarded a Bachelor of Arts by Research degree (the predecessor of the PhD degree) in March 1916 for his work on **highly composite numbers**, sections of the first part of which had been published the preceding year in the Proceedings of the London Mathematical Society. The paper was more than 50 pages long and proved various properties of such numbers. Hardy disliked this topic area but remarked that though it engaged with what he called the 'backwater of mathematics' in it, Ramanujan displayed 'extraordinary mastery over the algebra of inequalities'.

In 1916, Ramanujan's tutor Barnes had written to the University of Madras that, given Ramanujan's achievements, it seemed likely he would be elected a **Fellow of Trinity College** the following October. But October 1917 came and went without Ramanujan's election. At the time, the college was wracked with dissension over the Bertrand Russell affair, and Ramanujan's champion, Hardy, was squarely in the out-of-favour camp. Then, too, it seems certain, in light of future events, simple **racism was a factor**; Ramanujan, after all, was a black man. The disappointment left Ramanujan's mood darker, the whole structure of his personality that much shakier.

3.2 RAMANUJAN IN MATLOCK SANATORIUM FOR THE TREATMENT OF T.B.

It was around this time that he entered **Matlock*,** which could hardly have lifted his spirits. English sanatoriums were typically presided over by stern, patriarchal figures, strict disciplinarians who ruled with an iron hand. And Matlock was in the mould. A friend would later recall that Ramanujan was "cowed down by Dr. Ram, who seems to have told him, 'As long as you are a patient and not well you are not free and the doctor has control over your movements.' "

Matlock House Sanitorium, located in Derbyshire, England, was a 20th century health facility where Ramanujan stayed from late 1917 to mid-1918 for the treatment of tuberculosis.

Matlock was typical of English sanatoriums in at least one other respect: it was out-of-the-way geographically. "There is much evidence to suggest," writes **Bryder** in her study of British tuberculosis care, "that inmates of sanatoria felt estranged from the outside world. Not only did their geographical isolation make visits difficult and therefore infrequent, but social attitudes accentuated that isolation." At Matlock, located about 150 miles northwest of London, in the Peak District of Derbyshire, Ramanujan enjoyed no steady stream of visitors. Getting there was not easy. The following year, a friend who did make the trip, A. S. Ramalingam, would write of the "cold weary journey" on the night train he'd had to endure. Hardy was almost certainly thinking of the Matlock period when he wrote, a few months after Ramalingam's visit, of Ramanujan's "long illness and the spells of comparative solitude" influencing his mental state. At Matlock, Ramanujan was cold and miserable much of the time. The Trinity fellowship rejection rankled. And he was too sick to be productive mathematically, which also distressed him. The doctors didn't suit him. He couldn't get food to eat and the food he could get he didn't like; sometimes he craved the hot **dosai**, a kind of pancake, that Anantharaman's mother would cook up for him back in Kumbakonam. Finally, he was getting little in the way of nurturing and emotional support from home. And he wasn't getting it from Hardy, either.

He grew profoundly depressed. At One point, he had nightmares in which he was visited by images of his own abdomen as a kind of mathematical appendage with "singularities," points in space marked by indefinable mathematical surges like those he and Hardy had explored in their partitions work. Intense pain might show up at x = I, half as much pain at x = - I, and so on. The nightmares recurred. Ramanujan was at a low ebb, balanced precariously on the edge of **mental instability**.

Sometime in early 1917 something had gone badly wrong with Ramanujan's body; he had come down sick. But by the end of the year, it was not only his body that was troubled, but his mind. By then, certainly, he was not a happy man. There were some disturbing developments at his home too.

Ramanujan got no letters from his wife Janaki not because she didn't write to Ramanujan but because his mother intercepted those letters. One time, with a package destined for Ramanujan prepared for pickup and her mother-in-law out of the house, Janaki slipped into it a brief note. But Komalatammal returned early, spied the note, opened it, read it, dismissed it as childish or silly or stupid, and refused to send it. Janaki was upset. But what could she say? Or do? She was a girl of seventeen, and her mother-in-law was, well ... her mother-in-law. So, she said nothing, consoling herself that that was just how things were. That was just how things were in the Indian extended family. Mother-in-law and wife troubles were common in many Indian households, perhaps most. They were the stuff of jokes-and, as usual, the butt of moralistic opprobrium from Western observers, whose sensibilities were offended by so much of what they saw in India. The institution of child-brides, was abomination enough, but the girl's customary fate in her new home was, if anything, even worse. It is pitiable for the child-wife, torn from a home that contained all she knew of happiness, to be obliged to submit herself to the temper, caprice, and often tyranny of her husband, but when to this is added the despotism and cruelty of several elderly women, who often avail themselves of her helplessness, and if she fails to find favour in her husband's eyes, almost invariably take their cue of unkind conduct from him, her lot may be better imagined than described. She has absolutely no place to go to for comfort and sympathy if it is not to be found in her new home. There is no escape, and no matter what her sufferings, her parents' home is closed to her. An appeal to them meets with a rigid command to submit herself to her husband.

There were seeds of domestic tension in such a situation. And in the case of Ramanujan's family, they grew in more than usually fertile soil. For one thing, Komalatammal's identification with her son-his birth long sought and the object of fervent prayers-was particularly strong. And while young Indian men were often more attached to mother than to wife, Ramanujan was more than usually so; he looked like her, thought like her, shared the ardency of her temperament. Then, too, Ramanujan's mother, Komalatammal, close to fifty by now, was a formidable figure of a woman, by one account "insanely jealous" of her son, while his wife Janaki was a meek and callow teenager.

Ramanujan's mother Komalatammal's side of the conflict does not come down to us, except that, by some accounts, she blamed her daughter-in law Janaki, on the basis of her horoscope, for Ramanujan's ill health; had he married someone else, she was certain, he would not have gotten sick. It may be, however, that she bore toward Janaki no special animosity at all, but merely wrote her off as the child she still was. Almost forty years her senior and used to having her way around the house-and with Ramanujan-she could scarcely be expected to give Janaki much voice or autonomy until she grew up. It demands little imagination, then, to see her dismissing with scarcely a wave of her hand Janaki's pleas and protestations.

Finally, Janaki found an excuse to get away. Her only brother, Srinivasa Iyengar, then working in Karachi, in what is now Pakistan, was getting married. The wedding would be back in her hometown of Rajendram, where she and Ramanujan had been wed. Obviously she would have to go. Through mutual friends in Kumbakonam, her parents had known something was amiss, and it may have been their idea to use the wedding as pretext to get Janaki out of Komalatammal's clutches.

A little later, Janaki, now at her brother's house in Karachi, wrote to Ramanujan, and this time the letter got through. "Could he send her some money for a new sari and for a wedding gift for her brother?" Dutifully, Ramanujan sent the money. But by now bitter at the long silence from his wife and knowing only what he heard from his mother, he let no warmth or feeling slip into his reply. The trouble at home had overflowed its banks, distorting Ramanujan's relationship with his whole family. His letters home first dropped off, then stopped altogether. In 1914, Ramanujan had written home three or four times a month. By 1916, sometimes two or three months passed before he wrote. During 1917, the family heard from him not at all. For a long time, perhaps shamed at feeling abandoned by his own family, Ramanujan kept silent about it. But finally, while in the Cambridge nursing home, he could hold it in no longer. He told Hardy. And he told his friend Chatterji. Visiting him one day, Chatterji found him looking unhappy. "What's the matter?" he asked. "Oh, my house has not written to me," Ramanujan replied, using a common South Indian idiom for "my wife." "Well," joked Chatterji, though familiar with the idiom, "houses don't write."

The "quarrel" disturbed Ramanujan's equanimity, made for a tangled snag in his emotional lifeline back to India. That was bad enough. But that Hardy knew so little of his personal side at this late date, three and a half years after he came to England, testified to something more-and, perhaps, worse. Presumably, Hardy was his best friend in England. At least before he got sick, the two had seen each other almost every day.

On February II, 2018, ending a period of silence that had extended over more than a year, Ramanujan wrote to his family in India. About this same time, perhaps a little earlier, during a period when he was briefly away from Matlock, **he tried to kill himself**. One day in January or February of 1918, it was at a station somewhere in this network, then smaller and newer, that **Ramanujan threw himself onto the tracks in front of an approaching train**. What happened next would be easy enough to read as a miracle. A guard spotted him do it and pulled a switch, bringing the train screeching to a stop a few feet in front of him. **Ramanujan was alive,** though bloodied enough to leave his shins deeply scarred. **He was arrested and hauled off to Scotland Yard.** Called to the scene, Hardy, marshalling all his charm and academic stature, made a show of how there, before the police, stood the great Mr. Srinivasa Ramanujan, a Fellow of the Royal Society, and how a Fellow of the Royal Society simply could not be arrested.

In fact, Ramanujan was not an F.R.S. He would hardly have been immune from arrest in any case, and the police were not fooled for a minute. But they investigated, learned Ramanujan was indeed reputed to be an eminent mathematician, and decided to let him go. "We in Scotland Yard did not want to spoil [his] life," the officer in charge of the case said later.

3.3 RAMANUJAN ELECTED AS A FELLOW OF THE ROYAL SOCIETY (F.R.S.) IN 1918

On 6 December 1917, Ramanujan was elected to the **London Mathematical Socie**ty. **On 2 May 1918, he was elected a Fellow of the Royal Society,** the second Indian admitted, after Ardaseer Cursetjee in 1841. At the age of 31, Ramanujan was one of the youngest Fellows in the Royal Society's history. He was elected "for his investigation in **elliptic functions and the Theory of Numbers**." On 13 October 1918, he was the first Indian to be elected a **Fellow of Trinity College, Cambridge.**

India, which soon heard the news, was thrilled. The Madras members of the Indian Mathematical Society wrote Hardy in thanks "for the aid and guidance you have been extending to Mr. S. Ramanujan in his work." In a postscript, P. V. Seshu Iyer added in his letter to Hardy, extending personal thanks for the "care you have been bestowing on him during these months when his health has not been good."

Among the colleges of Cambridge University, **Trinity** was the largest, with the most lustrous heritage, home to kings, poets, geniuses. **Isaac Newton himself had studied there**; since 1755, his marble likeness, holding the prism he'd used to explore the polychromatic nature of light, stood in its chapel. Lord Byron had gone to Trinity. So had Tennyson, Thackeray, and Fitzgerald. So had the **historian Macaulay**, and the **physicist Rutherford**, and the philosopher **Bertrand Russell**. So had five English prime ministers. **And now, Ramanujan was at Trinity, too.**

3.4 ILLNESS AND DEATH OF RAMANUJAN

Ramanujan had numerous health problems throughout his life. His health worsened in England; possibly he was also less resilient due to the difficulty of keeping to the strict dietary requirements of his religion there and because of wartime rationing in 1914–1918. He was diagnosed with **tuberculosis** and a **severe vitamin deficiency**, and confined to a sanatorium. **He attempted suicide** in late 1917 or early 1918 by jumping on the tracks of a London underground station. **Scotland Yard arrested him for attempting suicide** (which was a crime), but released him after Hardy intervened.

In 1919, Ramanujan returned to Kumbakonam, Madras Presidency, where he died on 26 April 1920 aged 32.

"Where is she?" asked Ramanujan of his mother as he stepped off the ship in Bombay on March 27,1919. 'She' was Janaki. His mother was there, and his brother Lakshmi Narasimhan; the two of them had set out for Bombay on the twenty-first. But not his wife. Why fret over Janaki? sniped Komalatammal. Scarcely off the boat, Ramanujan had dropped into the family snake pit. Domestic conflict had sabotaged his last three years in England. Now they dampened his arrival in India. In fact, the two sides of the family had been out of touch for more than a year. No one knew just where Janaki was. Perhaps back in Rajendram. Or, if she had learned of Ramanujan's arrival, may be with her sister in Madras. So, Lakshmi Narasimhan was dispatched to write two identical letters to Janaki asking that she come meet Ramanujan in Madras.

Janaki was back in Rajendram, it turns out, where she'd gone for her brother's wedding more than a year before. She did know of Ramanujan's return-but only thanks to the Madras papers, not because her mother-in-law had bothered to inform her. Her brother, R. Srinivasa Iyengar, had advised her not to rejoin the family in Madras. Komalatammal hated her; why place herself within the tiger's jaws once again? But then the letter came from Lakshmi Narasimhan. Ramanujan, it said, wanted her. That was all it took. She and her brother set off. "When I go back I shall never be asked to a funeral," Ramanujan had told Neville before he left. It was one form of the taint he expected to bear, in the eyes of strictly orthodox Brahmins, for going to England. To help remove it, Komalatammal had planned to take her son directly to Rameswaram, for purification ceremonies in the great temple to which the whole family had earlier gone for a pilgrimage. But now, in Bombay, one look at him decided her against it; Rameswaram, all the way south almost to Ceylon, was another five hundred miles and another hard day on the train beyond Madras. He was just too sick. And so, after a few nights in Bombay, Ramanujan, his mother, and brother boarded the Bombay Mail for the overnight trip to Madras.

on January 12, 1920, a couple of months before his death in April 1920, Ramanujan wrote to Hardy for the first time in almost a year.

I am extremely sorry for not writing to you a single letter up to now.... I discovered very interesting functions recently which I **call "Mock" theta functions. Unlike the "False" theta functions** (studied partially by Prof. Rogers in his interesting paper) they enter into mathematics as beautifully as the ordinary theta functions. I am sending you with this letter some examples.

In how it would excite the active interest of mathematicians up until the present, it was a letter much like the one Ramanujan had written Hardy seven years earlier almost to the day. By one estimate, it represented "one of the most original pieces of mathematics, and in some ways the very best, which Ramanujan did." A mathematician, Watson wrote:

"Ramanujan's discovery of the mock-theta functions makes it obvious that his skill and ingenuity did not desert him at the oncoming of his untimely end. As much as any of his earlier work, the mock-theta functions are an achievement sufficient to cause his name to be held in lasting remembrance. To his students such discoveries will be a source-of-delight and wonder until the time shall come when we too shall make our journey to that Garden of Proserpine where:

"In Somewhat Indifferent Health" Pale, beyond porch and portal Crowned with calm leaves, she stands Who gathers all things mortal With cold, immortal hands."

After his death, his brother Tirunarayanan compiled Ramanujan's remaining handwritten notes, consisting of formulae on singular moduli, hypergeometric series and continued fractions. In his last days, though in severe pain, "he continued doing his mathematics filling sheet after sheet with numbers", Ramanujan's widow, Janaki Ammal, recounts.

After Ramanujan's death, Janaki Ammal (his widow) moved to Bombay. In 1931; she returned to Madras and settled in Triplicane, where she supported herself on a pension from Madras University and income from tailoring. In 1950, she adopted a son, **W. Narayanan**, who eventually became an officer of the State Bank of India and raised a family. In her later years, she was granted a lifetime pension from Ramanujan's former employer, the Madras Port Trust, and pensions from, among others, the Indian National Science Academy and the state governments of Tamil Nadu, Andhra Pradesh and West Bengal. She continued to cherish Ramanujan's memory, and was active in efforts to increase his public recognition; prominent mathematicians, including George Andrews, Bruce C. Berndt and Béla Bollobás made it a point to visit her while in India. She died at her Triplicane residence in 1994.

A 1994 analysis of Ramanujan's medical records and symptoms by D. A. B. Young concluded that his medical symptoms—including his past relapses, fevers, and hepatic conditions—were **much closer to those of hepatic amoebiasis,** an illness then widespread in Madras, **than of tuberculosis**. He had two episodes of dysentery before he left India. When not properly treated, amoebic dysentery can lie dormant for years and lead to hepatic amoebiasis, whose diagnosis was not then well established. At the time, if properly diagnosed, amoebiasis was a treatable and often curable disease; British soldiers who contracted it during the First

World War were being successfully cured of amoebiasis around the time Ramanujan left England.

3.5 PERSONALITY AND SPIRITUAL LIFE OF RAMANUJAN

While asleep, I had an unusual experience. There was a red screen formed by flowing blood, as it were. I was observing it. Suddenly a hand began to write on the screen. I became all attention. That hand wrote a number of elliptic integrals. They stuck to my mind. As soon as I woke up, I committed them to writing.

—Srinivasa Ramanujan

Ramanujan has been described as a person of a somewhat shy and quiet disposition, a dignified man with pleasant manners. He lived a simple life at Cambridge. Ramanujan's first Indian biographers describe him as a rigorously orthodox Hindu. He credited his acumen to his family goddess, **Namagiri** Thayar (Goddess Mahalakshmi) of Namakkal. He looked to her for inspiration in his work and said he dreamed of blood drops that symbolised her consort, Narasimha. Later he had visions of scrolls of complex mathematical content unfolding before his eyes. He often said, "An equation for me has no meaning unless it expresses a thought of God."

Hardy cites Ramanujan as remarking that all religions seemed equally true to him. Hardy further argued that Ramanujan's religious belief had been romanticised by Westerners and overstated—in reference to his belief, not practice—by Indian biographers. At the same time, he remarked on Ramanujan's strict vegetarianism.

Similarly, in an interview with Frontline, Berndt said, "Many people falsely promulgate mystical powers to Ramanujan's mathematical thinking. It is not true. He has meticulously recorded every result in his three notebooks," further speculating that Ramanujan worked out intermediate results **on slate because he could not afford the paper to record them more permanently.**

Berndt reported that Janaki said in 1984 that Ramanujan spent so much of his time on mathematics that he did not go to the temple, that she and her mother often fed him because he had no time to eat, and that most of the religious stories attributed to him originated with others. However, his orthopraxy was not in doubt.

3.6 ANECDOTE OF PROF. P.C. MAHALANOBIS

"I have mentioned that Ramanujan and I often used to go out for long walks on Sunday mornings. During these walks our discussions ranged over a wide variety of subjects. He had some progressive ideas about life and society but no reformist views. Left to himself, he would often speak of certain philosophical questions. He was eager to work out a theory of reality which would be based on the fundamental concepts of "zero", "infinity" and the set of finite numbers. I used to follow in a general way but I never clearly understood what he had in mind. He sometimes spoke of "zero" as the symbol of the absolute (Nirguna-Brahmam) of the extreme monistic school of Hindu philosophy, that is, the reality to which no qualities can be attributed, which cannot be defined or described by words, and which is completely beyond the reach of the human mind. According to Ramanujan, the appropriate symbol was the number "zero", which is the absolute negation of all attributes. He looked on the number "infinity" as the totality of all possibilities, which was capable of becoming manifest in reality and which was inexhaustible. According to Ramanujan, the product of infinity and zero would supply the whole set of finite numbers. Each act of creation, as far as I could understand, could be symbolized as a particular product of infinity and zero, and from each such product would emerge a particular individual of which the appropriate symbol was a particular finite number. I have put down what I remember of his views. I do not know the exact implication. He seemed to have been perhaps emotionally more interested in his philosophical ideas than in his mathematical work. He spoke with such enthusiasm about the philosophical questions that sometimes I felt he would have been better pleased to have succeeded in establishing his philosophical theories than in supplying rigorous proofs of his mathematical conjectures."

"Ramanujan had a somewhat shy and quiet disposition, a dignified bearing, and pleasant manners. He would listen carefully to that other people were saying but would usually remain silent. If he was asked any question, or on rare occasions, if he joined in any general conversation, he would speak frankly, but briefly. Whilst speaking to a friend or in very small groups, he would, however, expound his own ideas with great enthusiasm, not only on philosophical questions but occasionally also on other subjects in which he was seriously interested. Although I could not follow his mathematics, he left a lasting impression on my mind. His bright eyes and gentle face with a friendly smile are still vivid in my mind."

Mr. Jawaharlal Nehru wrote the following in his book 'Discovery of India' about Ramanujan in 1946:

"Ramanujan's brief life and death are symbolic of conditions in India. Of our millions how few get any education at all; how many live on the verge of starvation ... If life opened its gates to them and offered them food and healthy conditions of living and education and opportunities of growth, how many among these millions would be eminent scientists, educationists, technicians, industrialists, writers, and artists, helping to build a new India and a new world?"

A writer for the Illustrated Weekly of India, referring to Ramanujan, suggested that "*perhaps the luminescence of his mind is too harsh for Indians in this age of their intellectual bleakness....*"

Ramanujan's emergence as a figure on the world's mathematical stage was given in an article appearing in India during the centennial, by another writer for The Illustrated Week(y of India known only as "RGK.

"Ramanujan was svayambhu-"self-born." He had sprouted up out of the soil of India of his own accord. He had created himself.

In India, particularly in South India today, everyone has heard of Ramanujan. College professors and rickshaw drivers alike know his story, at least in sketchy outline, just as everyone in the West knows of Einstein. Few can say much about his work, and yet something in the story of his struggle for the chance to pursue his work on his own terms compels the imagination, leaving Ramanujan a symbol for genius, for the obstacles it faces, for the burdens it bears, for the pleasure it takes in its own existence.

SECTION-2

SIGNIFICANT CONTRIBUTIONS OF S. RAMANUJAN

CHAPTER 4

Mathematical Contributions of Ramanujan

Ramanujan's mathematics remains youthful even in the modern world of the computer. His modular equations were used by Canadians Jonathan and Peter Borwein to calculate π (the ratio of the circumference of a circle to its diameter) to several million decimal places. Borwein showed that these modular equations produce efficient algorithms to obtain approximations to π and other numbers.

More recently, Ramanujan's transformations for elliptic functions were used by David and Gregory Chudnovsky to produce very rapidly convergent algorithms to compute π; in fact, the Chudnovskys have now calculated π to the order of about a billion digits!

Ramanujan has left behind enough ideas to keep mathematicians busy well into the twenty-first century. Professor Dyson has remarked that we should be grateful to Ramanujan not only for discovering so much but also for providing others plenty to discover! What sets Ramanujan apart from the rest of the mathematical giants is that feeling of astonishment he creates with his stunningly beautiful identities. Ramanujan is like a gem with many faces. His identities can be studied from different viewpoints. Each face of this gem dazzles the beholder with its array of colours!

4.1 SOME CONTRIBUTIONS OF RAMANUJAN

Despite lacking formal training, Ramanujan independently discovered thousands of results, including the partition function, mock theta functions, and Ramanujan prime/summation, profoundly impacting 20th-century mathematics.

Ramanujan's contribution extends to mathematical fields such as complex analysis, number theory, infinite series, and continued fractions.

Infinite series for pi: In 1914, Ramanujan found a formula for *infinite series for pi,* which forms the basis of many algorithms used today. Finding an *accurate approximation of π (pi)* has been one of the most important challenges in the history of mathematics.

Game theory: Ramanujan discovered a long list of new ideas for solving many challenging mathematical problems that have given great impetus to the development of game theory. His contribution to game theory is purely based on intuition and natural talent and is unmatched to this day.

Mock theta function: He elaborated on the mock theta function, a concept in the field of modular forms of mathematics.

Ramanujan number: 1729 is known as the Ramanujan number which is the sum of the cubes of two numbers 10 and 9.

Circle Method: Ramanujan, along with GH Hardy, invented the circle method which gave the first approximations of the partition of numbers beyond 200. This method contributed significantly to solving the notorious complex problems of the 20th century, such as Waring's conjecture and other additional questions.

Theta Function: Theta function is a special function of several complex variables. German mathematician Carl Gustav Jacob Jacobi invented several closely related theta functions known as Jacobi theta functions. Theta function was studied by extensively Ramanujan who came up with the Ramanujan theta function, that generalizes the form of Jacobi theta functions and also captures general properties. Ramanujan theta function is used to determine the critical dimensions in Bosonic string theory, superstring theory, and M-theory.

Other notable contributions by Ramanujan include hypergeometric series, the Riemann series, the elliptic integrals, the theory of divergent series, and the functional equations of the zeta function.

Ramanujan's achievements were all about elegance, depth, and surprise beautifully intertwined. Unfortunately, Ramanujan contracted a fatal illness in England in 1918. He convalesced there for more than a year and returned to India in 1919. His condition then worsened, and he died on 26 April 1920. One might expect that a dying man would stop working and await his fate. However, Ramanujan spent his last year producing some of his most profound mathematics.

It has been more than a century however; his mathematical discoveries are still alive and flourishing. "Ramanujan is important not just as a mathematician but because of what he tells us that the human mind can do". "Someone with his ability is so rare and so precious that we can't afford to lose them. A genius can arise anywhere in the world. It is our good fortune that he was one of us. It is unfortunate that too little of Ramanujan's life and work, esoteric though the latter is, seems to be known to most of us".

4.1.1 Key contributions of Ramanujan include:

(i) **Number Theory & Partitions:** He found deep, unexpected properties of the partition function (representing a number as a sum of positive integers), developing the Ramanujan-Hardy asymptotic formula.

(ii) **Infinite Series** He derived numerous rapidly converging infinite series for calculating to high precision.

(iii) **Continued Fractions & Modular Functions:** He developed new techniques for analysing Continued Fractions, and discovered modular forms that are still used today.

(iv) **Lost Notebook & Mock Theta Functions:** Towards the end of his life, he developed "mock theta functions," which are functions that have applications in black hole physics today.

(v) **Number 1729:** He is famous for identifying 1729 as the smallest number expressible as the sum of two cubes in two different ways (Hardy-Ramanujan number).

4.2 PAUL ERDÖS AND RAMANUJAN

Paul Erdös, a distinguished member of the Hungarian Academy of Sciences, told 'The Hindu' in Gainesville, Florida, USA, that he was greatly inspired by Ramanujan's work. He said that Ramanujan's contribution to number theory in collaboration with Hardy formed the basis for his work which led to the creation of probabilistic number theory.

The passing away of Paul Erdös in 1996 marked the end of a great era. Erdös was the most prolific mathematician in history and influenced the academic lives of many including me. His path-breaking work on probabilistic number theory was inspired by the seminal 1917 paper of Hardy and Ramanujan.

Paul Erdös said that when Hardy was asked what was his greatest contribution to mathematics, **he unhesitatingly said "The discovery of Ramanujan".** Hardy told him that Ramanujan went far beyond his theorems. Hardy once gave an estimation of mathematicians on the basis of pure talent on a scale of 1–100. Professor Erdös said that Hardy gave Ramanujan 100, 80 to the famous mathematician David Hilbert, 30 to colleague Littlewood, and only 25 to himself. "Although Hardy was modest in giving himself only 25, the fact that he gave 100 to Ramanujan revealed the regard he had for Ramanujan's work." A man who lived by numbers, Prof. Erdös noted that Ramanujan was so multifaceted that no mathematician could fully decipher or comprehend all of his creations. "I wish I had a chance to meet Ramanujan. Unfortunately, he died when I was seven." Hardy was of the opinion that education would not have made Ramanujan a greater mathematician; it might have stifled his genius. However, Professor Erdös expressed the view that education would have proved a great deal more for Ramanujan. "He would not have wasted so much time rediscovering the work of other mathematicians."

In September 1996, Professor Paul Erdös, one of the mathematical legends of this century, died while attending a conference in Warsaw, Poland. His death at the age of 83 marked the end of a great era, for Erdös was not only an outstanding mathematician but a very kind and generous human being who encouraged hundreds of mathematicians over the decades, especially young aspirants to the subject. Many owe their careers to him. He was without doubt the most prolific mathematician of this century, having written more than 1000 papers, a significant proportion of them being joint papers. Even in a mathematical world which is used to geniuses and their idiosyncrasies, Erdös was considered an unusual phenomenon and was viewed with awe and adoration just as Ramanujan evoked surprise and admiration. **And like Ramanujan's mathematics, the contributions of Erdös will continue to inspire and influence research in the decades ahead.**

4.3 RAMANUJAN'S MAGIC SQUARE

The origin of the Chapter on "Magie Squares' probably is found in Ramanujan's early school days and is therefore much earlier than the remainder of the notebooks. Rules for constructing certain rectangular arrays of natural numbers are given. Most of Ramanujan's attention is devoted to constructing the magic squares.

A magic square is a square array of (usually distinct) natural numbers so that the sum of the numbers in each row, column, or diagonal is the same. In some instances, the requirement on the two diagonal sums is dropped. In the notebooks, Ramanujan uses the word "corner" for "diagonal." We emphasize that the theory of magic squares is barely begun by Ramanujan in this chapter.

Considerably more extensive developments are contained in other books. Ramanujan commences this Chapter with the following simple principle for constructing magic squares

Consider two sets of natural numbers S, = {A, B, C, . ..} and S, = (P, Q, R, . ..}. each with n elements. Take the n_2 numbers in the direct sum S, + S, and arrange them in an (n x n) square so that each letter appears exactly once in each row, column, and diagonal. Clearly, we have then constructed a magic square. Of course, some numbers may appear more than once.

In Corollary 1, Ramanujan states the trivial fact that if A + P, A + Q, A + R, . . . are in arithmetic progression, then B + P, B + Q, B + R, . . . are also in arithmetic progression.

In Corollary 2, Ramanujan remarks that if A + P, A + Q, A + R, A + S, . . . are known and also B + P is given, then we can determine B + Q, B + R, B + S, ... This is clear, for B-A is thus known, and we may Write B + Q = (B - A) + (A + Q), etc. Ramanujan informs us that in constructing a magic square, we should not give values to A, B, C, . . and P, Q, R, ... but instead, values should be assigned to A+P, A+Q, A+R ,....The reason for this advice is not clear, for in either case 2n parameters need to be prescribed.

Example: Given that A+P=8, B+P=l0, C+P=11, D+P=14, and C+R=25, find A+R, B+R, find D+R.

Solution: Since R - P = 14, then A + R = (A + P) + (R -. P) = 22. Similarly, B+R=24 and D+R=28.

In mathematics, especially historical and recreational mathematics, a square array of numbers, usually positive integers, is called a magic square if the sums of the numbers in each row, each column, and both main diagonals are the same. The order of the magic square is the number of integers along one side (n), and the constant sum is called the magic constant. If the array includes just the positive integers :1, 2, ...$n^{2,}$ the magic square is said to be normal. Many authors take magic square to mean normal magic square.

Magic squares that include repeated entries do not fall under this definition and are referred to as trivial. Some well-known examples, including the Sagrada Família magic square are trivial in this sense. When all the rows and columns but not both diagonals sum to the magic constant, this gives a semi-magic square (sometimes called ortho-magic square).

The mathematical study of magic squares typically deals with its construction, classification, and enumeration. Although completely general methods for producing all the magic squares of all orders do not exist, historically three general techniques have been discovered: by bordering, by making composite magic squares, and by adding two preliminary squares. There are also more specific strategies like the continuous enumeration method that reproduces specific patterns. Magic squares are generally classified according to their order n as: odd if n is odd, evenly even (also referred to as "doubly even") if n is a multiple of 4, oddly even (also known as "singly even") if n is any other even number. This classification is based on different techniques required to construct odd, evenly even, and oddly even squares. Beside this, depending on further properties, magic squares are also classified as associative magic squares, pan-diagonal magic squares, most-perfect magic squares, and so on. More challengingly, attempts have also been made to classify all the magic squares of a given order as transformations of a smaller set of squares. Except for $n \leq 5$, the enumeration of higher-order magic squares is still an open challenge. The enumeration of most-perfect magic squares of any order was only accomplished in the late 20th century.

Magic squares have a long history, dating back to at least 190 BCE in China. At various times they have acquired occult or mythical significance, and have appeared as symbols in works of art. In modern times they have been generalized a number of ways, including using extra or different constraints, multiplying instead of adding cells, using alternate shapes or more than two dimensions, and replacing numbers with shapes and addition with geometric operations.

(i) 3x3 Magic Squares:

A 3x3 magic square using numbers 1–9 features a "magic constant" of 15, where every row, column, and diagonal sums to 15, with 5 always in the centre. The standard, common solution is:

8	1	6
3	5	7
4	9	2

Following are two more examples of 3x3 Magic squares, where sum of each row, column, and diagonal is constant.

10	2	7
4	6	9
5	11	3

14	5	12
7	11	13
10	15	6

(ii) 4x4 Magic Squares:

We study here below, a particular 4x4 magic square, commonly known as **"Ramanujan's magic**

square". Is there any "magic" in magic squares ? The Ramanujan "magic square" gives us a clue to this question. A well-crafted square matrix can tell us so many things and even show us some "magic". Let us take the date of birth of the legendary Srinivasa Ramanujan 1 , as an example (he was born on 22- 12-1887).

DD	MM	CC	YY
22	12	18	87

Now, create a 4x4 matrix as follows:

22	12	18	87
21	84	32	02
92	16	07	24
04	27	82	26

Notice the following magic in the above matrix (aka magic square):

- Each row adds up to **139** (e.g. 22 + 12 + 18 + 87)
- Each column adds up to **139** (e.g. 22 + 21 + 92 + 04)
- All the corners add up to **139** (e.g. 22 + 87 + 04 + 26)
- All the elements of the 2x2 submatrix in the middle add up to **139** (e.g. 84 + 16 + 32 + 07)
- All the elements of each diagonal add up to **139** (e.g. 22 +84 + 07 + 26).

Now, divide the 4x4 matrix into four 2x2 submatrices by drawing a vertical line in the middle and a horizontal line in the middle of the 4x4 matrix. Now, notice that:

- All the elements of each 2x2 submatrix add up to 139 (e.g. 22 + 12 + 21 + 84).

A closer look at the above 4x4 matrix will show that there is really no "magic" in the above magic square 2 . The above matrix is made with the following elements:

DD	MM	CC	YY
CC+3	YY-3	DD+10	MM-10
YY+5	CC-2	MM-5	DD+2
MM-8	DD+5	YY-5	CC+8

Using the same idea as above, you can create another magic square easily. All you need to do is to choose the matrix elements accordingly. Try this:

22	12	18	87
88	17	09	25
10	24	89	16
19	86	23	11

You can construct any number of "magic" squares like the above, if you follow the above logic. You can start with any set of four numbers for the top row, and fill the rest of the square with cleverly chosen numbers. In fact, you can build such "magic" squares for anybody's date-of-birth (even yours).

Let me try with my date of birth, which is: 01 June 1948. So, following is the 4x4 magic square for me:

01	06	19	48
22	45	11	-4
53	17	01	03
-2	06	43	27

Yo will notice that:

- Each row adds up to **74**.
- Each column adds up to **74**
- All the corners add up to **74**
- All the elements of the 2x2 submatrix in the middle add up to **74**
- All the elements of each diagonal add up to **74.**

4.4 SOME MORE MATHEMATICAL ACHIEVEMENTS OF RAMANUJAN

Ramanujan has to be measured only by the impact of his contributions. One may feel that the most important papers are those that solve longstanding problems. These are significant, but so also are those that trigger new questions and open up fresh avenues of thought. The great mathematician Bernhard Riemann, who worked mainly in analysis, wrote just one paper in number theory with the intention of proving the prime number theorem that was conjectured nearly a century earlier by Gauss. Although Riemann did not prove this theorem, he created a totally new approach and raised several questions that have kept mathematicians busy ever since. Even Ramanujan pursued some ideas akin to those in Riemann's paper. Certain questions raised in that paper have been solved, but one known as "The Riemann Hypothesis" remains unsolved to this day, and the very effort to find a solution has been rewarding.

In mathematics, there is a distinction between insight and formulating or working through a proof. Ramanujan proposed an abundance of formulae that could be investigated later in depth. G. H. Hardy said that Ramanujan's discoveries are unusually rich and that there is often more to them than initially meets the eye. As a byproduct of his work, new directions of research were opened up.

One of Ramanujan's remarkable capabilities was the rapid solution of problems, illustrated by the following anecdote about an incident in which P. C. Mahalanobis posed a problem:

Imagine that you are on a street with houses marked 1 through n. There is a house in between (x) such that the sum of the house numbers to the left of it equals the sum of the house numbers to its right. If n is between 50 and 500, what are n and x?' This is a bivariate problem with multiple solutions. Ramanujan thought about it and gave the answer with a twist: He gave a continued fraction. The unusual part was that it was the solution to the whole class of problems.

4.5 ROGERS–RAMANUJAN IDENTITIES

This pair of identities (discovered independently by Rogers in 1894 and Ramanujan around 1910) are considered among the most beautiful in mathematics. The combinatorial description of the first identity is that the number of partitions of an integer n into parts differing by at least two equals the number of partitions of n into parts which when divided by 5 leave remainder 1 or 4. The second identity has a similar description. The simplicity of the identities belies their depth. Several proofs have been given, but none can be considered simple or straightforward.

In an attempt to understand these identities, a rich theory has developed, concerning, on the one hand, partitions whose parts satisfy gap conditions and, on the other, partitions whose parts satisfy congruence conditions. For a quarter century beginning around 1960, considerable work has been done in this direction, especially by Professor Basil Gordon of the University of California, Los Angeles, and by Professors George Andrews and David Bressoud of Pennsylvania State University. Rogers actually found several elegant companions to the Rogers–Ramanujan identities. Fundamental discoveries always find applications eventually. In 1979, the Australian mathematical physicist Rodney Baxter showed that the Rogers– Ramanujan identities and these companions are the solutions to the Hard Hexagon Model in Statistical Mechanics.

For this work, Professor Baxter was awarded the **Boltzmann medal** of the American Physical Society. In recent years, **identities of the Rogers–Ramanujan type** have found more applications to problems in mathematical physics. Ramanujan considered the Rogers–Ramanujan identities as arising out of a continued fraction possessing a product representation. It was Ramanujan's insight to have realised the importance of this continued fraction in the theory of modular forms. This is only one of many continued fractions studied by Ramanujan, but perhaps the most appealing. Professor Bruce Berndt of The University of Illinois has analysed several continued fractions of Ramanujan. These continued fractions can be approached in various ways and offer a wide range of problems for exploration.

4.6 RAMANUJAN'S NOTEBOOKS

While still in Madras, Ramanujan recorded the bulk of his results in four notebooks of loose leaf paper. They were mostly written up without any derivations. This is probably the origin of the misapprehension that Ramanujan was unable to prove his results and simply thought up the final result directly. Mathematician Bruce C. Berndt, in his review of these notebooks and Ramanujan's work, says that Ramanujan most certainly was able to prove most of his results, but chose not to record the proofs in his notes.

This may have been for any number of reasons. **Since the paper was very expensive, Ramanujan did most of his work and perhaps his proofs on slate**, after which he transferred the final results to paper. At the time, slates were commonly used by mathematics students in the Madras Presidency. He was also quite likely to have been influenced by the style of G. S. Carr's book, which stated results without proofs. It is also possible that Ramanujan considered his work to be for his personal interest alone and therefore recorded only the results.

The first notebook has 351 pages with 16 somewhat organised chapters and some unorganised material. The second has 256 pages in 21 chapters and 100 unorganised pages, and the third 33 unorganised pages. The results in his notebooks inspired numerous papers by later mathematicians trying to prove what he had found. Hardy himself wrote papers exploring material from Ramanujan's work, as did G. N. Watson, B. M. Wilson, and Bruce Berndt.

In 1976, George Andrews rediscovered a fourth notebook with 87 unorganised pages, the so-called "lost notebook".

4.7 THE LOST NOTEBOOK OF RAMANUJAN

Ramanujan's discoveries in India after return from England shortly before his death are the beginnings of theories which are grander in design and greater in depth. In the context of more modern methods, we now realise the significance of this last work of his on mock theta functions from the contents of the ***Lost Notebook***. Ramanujan was definitely on the rise, and we can only imagine the heights he would have reached had he lived longer. Ramanujan has no parallel in the history of mathematics. His life story is astounding to his own countrymen and, therefore, much more so to people of a totally different cultural background. The story of the ***Lost Notebook*** is equally amazing. During the final few months when Ramanujan was bedridden in Madras, he used to ask his wife Janaki Ammal for a scrap of paper to write down various formulae. Ramanujan wrote a letter to Hardy during this period outlining his discovery of the mock theta functions. Shortly after he died, Janaki Ammal collected these sheets and sent them to Cambridge. Many of these sheets were of thin fool-scape paper intermingled with thicker quarto sheets like the ones Hardy liked; so, Ramanujan must have taken some of these sheets of paper from Cambridge to India. After Ramanujan's death, B.M. Wilson began the task of editing the original two notebooks that Ramanujan had maintained in India prior to his departure to England. Hardy asked G.N. Watson to sort out this last manuscript of Ramanujan mainly because it was more in line with Watson's investigations. After Watson's death this manuscript along with several of Watson's papers were despatched to the Trinity College Library. Why this manuscript was not separated and classified under Ramanujan's name is unclear. At any rate it was forgotten that this manuscript was among Watson's papers, and so it soon acquired the name "The Lost Notebook," the romantic implications of which seemed to fit well with the atmosphere surrounding Ramanujan. In 1976, George Andrews of the Pennsylvania State University, the world authority on partitions and q-hypergeometric series, was in Cambridge looking through the Watson section. Quite unexpectedly he noticed a manuscript with incredibly beautiful formulae, and one look at them convinced him that these could only be written down by Ramanujan and that this was indeed the Lost Notebook! Andrews has been researching on the contents of the Lost Notebook ever since and has supplied proofs of many of the outstanding formulae contained therein. **The Lost Notebook contains about 650 formulae of which about 60 percent are on q-series and mock theta functions.** Theta functions are extremely interesting mathematical objects and equally useful as well, especially in physics. The mock theta functions mimic the theta functions in a certain way but are more general objects. Ramanujan wrote down several fascinating formulae for these functions which have interesting interpretations in the theory of partitions. About 30 percent of **the Lost Notebook** deals with the important topic of modular equations. Ramanujan demonstrates his unusual genius here by producing remarkable connections between this topic and q-series—this was so unexpected that Andrews was shocked on seeing it for the first time! The remaining 10 percent of the Lost Notebook contains miscellaneous results on integrals, Dirichlet series, and asymptotic formulae. It is indeed appropriate that during the Ramanujan Centennial celebrations in Madras, the Prime Minister released the first published copy of The Lost Notebook and presented it to George Andrews.

4.8 SPECIAL FUNCTIONS IN MATHEMATICS BY RAMANUJAN

Ramanujan wrote down several beautiful formulae involving various special functions (like the Beta and Gamma functions). For the past two decades, Professor Richard Askey of the University of Wisconsin, with his students and co-workers, systematically studied q-analogues of various special functions. In the course of this study, many of Ramanujan's identities found in his original notebooks and in the Lost Notebook were extremely useful. Many of the q-analogues found by Askey and others are now finding important applications in Physics, through the idea of Quantum Groups.

4.9 RAMANUJAN–PETERSSON CONJECTURE

Although there are numerous statements that could have borne the name *Ramanujan conjecture,* one was highly influential in later work. In particular, the connection of this conjecture with conjectures of André Weil in algebraic geometry opened up new areas of research. That Ramanujan conjecture is an assertion on the size of the tau-function, which has a generating function as the discriminant modular form $\Delta(q)$, a typical cusp form in the theory of modular forms. It was finally proven in 1973, as a consequence of Pierre Deligne's proof of the Weil conjectures. The reduction step involved is complicated. Deligne won a Fields Medal in 1978 for that work.

In his paper "On certain arithmetical functions", Ramanujan defined the so-called delta-function, whose coefficients are called $\tau(n)$ (the Ramanujan tau function). He proved many congruences for these numbers, such as $\tau(p) \equiv 1 + p^{11} \bmod 691$ for primes p. This congruence (and others like it that Ramanujan proved) inspired Jean-Pierre Serre (1954 Fields Medallist) to conjecture that there is a theory of Galois representations that "explains" these congruences and more generally all modular forms. $\Delta(z)$ is the first example of a modular form to be studied in this way. Deligne (in his Fields Medal-winning work) proved Serre's conjecture. The proof of Fermat's Last Theorem proceeds by first reinterpreting elliptic curves and modular forms in terms of these Galois representations. Without this theory, there would be no proof of Fermat's Last Theorem.

4.10 MOCK THETA FUNCTIONS

Ramanujan's work on mock theta functions is considered to be one of his deepest contributions. These results were discovered by him just before he died, and he communicated them to Hardy in his last letter dated January 1920. A major portion of ***The Lost Notebook*** is devoted to mock theta functions. In his letter Ramanujan listed several mock theta functions of orders three, five, and seven. Hardy passed on to Professor G.N. Watson the task of analysing Ramanujan's mock theta identities. Watson wrote two papers on this topic, the first of which was his presidential address to 'The London Mathematical Society' entitled "The Final Problem: An Account of the Mock Theta Functions." Watson explained the choice of the title as follows: "I doubt whether a more suitable title could be found for it than used by John H. Watson, M.D., for what he imagined to be his final memoir on Sherlock Holmes." Watson's first paper (1936) dealt with mock theta functions of third order, and the second (1937) with those of fifth order. Watson did not consider the seventh-order functions, but these were investigated by Selberg in 1938. In the last two decades Professor George Andrews has analysed and explained combinatorially many of Ramanujan's mock theta identities. In collaboration with his former student Frank Garvan, Andrews was led to conjecture that some of Ramanujan's mock theta identities were equivalent to certain results on partitions (a partition of a positive integer n is a representation of n as a sum of positive integers not exceeding n). These were called the "**Mock Theta Conjectures**." These conjectures were settled by Dean Hickerson in 1989, after the Ramanujan Centenary. Just this year Andrews and Hickerson have completed the study of eleven identities of Ramanujan on sixth-order mock theta functions in *The Lost Notebook*. Another recent advance is the work of Henri Cohen who explained certain mock theta identities in the context of Algebraic Number Theory.

Ramanujan listed 17 examples of functions that he called mock theta functions. He also listed several other examples of the function in his notebook. Ramanujan used the term "theta function" for what today would be called a modular form. Ramanujan conjectured that his mock modular forms corresponded to the ordinary modular forms earlier identified by Carl Jacobi, and that both would wind up with similar outputs for roots of 1. Nobody at the time understood what Ramanujan was talking about.

Ramanujan even divided these examples into groups based on their 'order', a term he never defined. Despite these formidable challenges, a few mathematicians such as G. E. Andrews, L. Dragonette, A. Selberg, and Watson investigated Ramanujan's mock theta functions .

Despite the absence of a theory, or much less, just a simple useful definition of a mock theta function, these few early works bolstered the belief that Ramanujan had discovered something important.

Watson, in his own words: "Ramanujan's discovery of the mock theta functions makes it obvious that his skill and ingenuity did not desert him at the oncoming of his untimely end. As much as any of his earlier works, the mock theta functions are an achievement sufficient to cause his name to be held in lasting remembrance. To his students such discoveries will be a source of delight and wonder until the time shall come when we too shall make our journey to that Garden of Proserpine (a.k.a. Persephone)".

These 17 examples, together with five examples of mock theta from Ramanujan's notebook have been related to a number of subjects including Lie theory, Modular forms, and Polymer chemistry. Despite this flurry of activity, the essence of Ramanujan's theory remained a mystery. The puzzle of his last letter to Hardy, thanks to the 'lost notebook', had morphed into the enigmatic web of Ramanujan's 22 mock theta functions. This strongly suggested the existence of a theory, and it also demanded a solution.

A New Beginning

As Ken Ono said, "... Although Ramanujan's last works provided the first examples of such forms, his untimely death and the enigmatic nature of his writings resulted in a great mystery. We will never know how he came up with the mock theta functions. We certainly cannot pretend to know what he fully intended to do with them. However, it is clear that he understood that the mock theta functions would go on to play important roles in number theory."

Ramanujan's mock theta functions or mock modular forms is not only an important tool in analytic number theory, it is applied in several other areas of mathematics like, topological invariants analysis, Lie superalgebras —to name a few.

Ramanujan died before he could prove his theorem. But more than 90 years later, Ono and his team have proved that these functions indeed mimicked modular forms, but don't share their defining characteristics, such as super-symmetry. The expansion of mock modular forms helps physicists computing the entropy or level of disorder of black holes.

4.11 HARDY–RAMANUJAN TAXI NUMBER, 1729

The number 1729 is known as the Hardy–Ramanujan number after a famous visit by Hardy to see Ramanujan at a hospital. In Hardy's words:

"I remember once going to see him when he was ill at Putney. I had ridden in taxi cab number 1729 and remarked that the number seemed to me rather a dull one, and that I hoped it was not an unfavourable omen." "No", he replied, "it is a very interesting number; it is the smallest number expressible as the sum of two cubes in two different ways."

Immediately before this anecdote, Hardy quoted Littlewood as saying, "Every positive integer was one of [Ramanujan's] personal friends."

The two different ways are:

$$1729 = 1^3 + 12^3 = 9^3 + 10^3.$$

Generalisations of this idea have created the notion of "taxi-cab numbers".

Ramanujan was mainly interested in equations which had solutions, and especially in providing algorithms or formulas for the solutions. This could be a reason why he was not interested in Fermat's Last Theorem. The famous Ramanujan taxi-cab number 1729 is an example of an integer which is expressible as a sum of cubes in two different ways (1729 is the sum of the cubes of 12 and 1, as well as the sum of the cubes of 10 and 9) and indeed is the smallest such example. One could view this problem as a mild variation of the one Fermat considered, namely, we are considering four non-zero integers x, y, z, w such that the sum of the cubes of x and y equals the sum of the cubes of z and w. Amazingly, addition of an extra variable w yields a solution. Euler had provided a general formula for solving this taxi-cab equation, but in one of Ramanujan's notebooks, there is also a general formula for the solution to this equation which is some ways is more elegant than Euler's.

4.12 RAMANUJAN'S SUM

In number theory, **Ramanujan's sum**, usually denoted $c_q(n)$, is a function of two positive integer variables q and n defined by the formula:

$$c_q(n) = \sum_{\substack{1 \le a \le q \\ (a,q)=1}} e^{2\pi i \frac{a}{q} n},$$

where (a, q) = 1 means that 'a' only takes on values coprime to q.

Srinivasa Ramanujan mentioned the sums in a 1918 paper. In addition to the expansions discussed in this article, Ramanujan's sums are used in the proof of Vinogradov's theorem that every sufficiently large odd number is the sum of three primes.

For integers *a* and *b*, $\frac{a}{b}$is read "*a* divides *b*" and means that there is

an integer *c* such that $\frac{b}{a} = c$.

Similarly, $a \nmid b$is read "*a* does not divide *b*".

The summation symbol

$$\sum_{d|m} f(d)$$

d goes through all the positive divisors of *m*, e.g.

$$\sum_{d|12} f(d) = f(1) + f(2) + f(3) + f(4) + f(6) + f(12).$$

$(a, b)$is the greatest common divisor,

ϕ (n) is Euler's totient function,

$\mu(n)$ is the Möbius function, and

$\zeta(s)$ is the Riemann zeta function.

Formulas for $c_q(n)$

These formulas come from the definition, Euler's formula $e^{ix} = \cos x + \mathrm{i}\sin x$, and elementary trigonometric identities.

$$\begin{aligned}
c_1(n) &= 1 \\
c_2(n) &= \cos n\pi \\
c_3(n) &= 2\cos\tfrac{2}{3}n\pi \\
c_4(n) &= 2\cos\tfrac{1}{2}n\pi \\
c_5(n) &= 2\cos\tfrac{2}{5}n\pi + 2\cos\tfrac{4}{5}n\pi \\
c_6(n) &= 2\cos\tfrac{1}{3}n\pi \\
c_7(n) &= 2\cos\tfrac{2}{7}n\pi + 2\cos\tfrac{4}{7}n\pi + 2\cos\tfrac{6}{7}n\pi \\
c_8(n) &= 2\cos\tfrac{1}{4}n\pi + 2\cos\tfrac{3}{4}n\pi \\
c_9(n) &= 2\cos\tfrac{2}{9}n\pi + 2\cos\tfrac{4}{9}n\pi + 2\cos\tfrac{8}{9}n\pi \\
c_{10}(n) &= 2\cos\tfrac{1}{5}n\pi + 2\cos\tfrac{3}{5}n\pi
\end{aligned}$$

and so, on $c_q(n)$ is always an integer.

Let $\zeta_q = e^{\frac{2\pi i}{q}}$.Then ζ_q is a root of the equation $x^q - 1 = 0$. Each of its powers,

$$\zeta_q, \zeta_q^2, \dots, \zeta_q^{q-1}, \zeta_q^q = \zeta_q^0 = 1$$

is also a root. Therefore, since there are q of them, they are all of the roots. The numbers ζ_q^n where $1 \le n \le q$ are called the q-th roots of unity. ζ_q is called a **primitive** q-th root of unity because the smallest value of n that makes $\zeta_q^n = 1$, is q. The other primitive q-th roots of unity are the numbers $\zeta_q^a$where $(a, q) = 1$.

Therefore, there are $\phi(q)$ primitive q-th roots of unity.

Thus, the Ramanujan sum $c_q(n)$ is the sum of the n-th powers of the primitive q-th roots of unity.

It is a fact that the powers of ζ_q are precisely the primitive roots for all the divisors of q.

Example. Let $q = 12$. Then

$\zeta_{12}, \zeta_{12}^5, \zeta_{12}^7$, and $\zeta_{12}^{11}$are the primitive twelfth roots of unity,

ζ_{12}^2 and $\zeta_{12}^{10}$are the primitive sixth roots of unity,

$\zeta_{12}^3 = i$ and $\zeta_{12}^9 = -i$are the primitive fourth roots of unity,

ζ_{12}^4 and $\zeta_{12}^8$are the primitive third roots of unity,

$\zeta_{12}^6 = -1$ is the primitive second root of unity, and

$\zeta_{12}^{12} = 1$is the primitive first root of unity.

Therefore, if

$$\eta_q(n) = \sum_{k=1}^{q} \zeta_q^{kn}$$

is the sum of the n-th powers of all the roots, primitive and imprimitive,

$$\eta_q(n) = \sum_{d|q} c_d\,(n),$$

Möbius inversion,

$$c_q(n) = \sum_{d|q} \mu\left(\frac{q}{d}\right) \eta_d(n).$$

It follows from the identity $x^q - 1 = (x - 1)(x^{q-1} + x^{q-2} + ... + x + 1)$ that

$$\eta_q(n) = \begin{cases} 0 & q \nmid n \\ q & q \mid n \end{cases}$$

$$c_q(n) = \sum_{d|(q,n)} \mu\left(\frac{q}{d}\right) d,$$

as published by Kluyver in 1906.

This shows that $c_q(n)$ is always an integer. Compare it with the formula

$$\phi(q) = \sum_{d|q} \mu\left(\frac{q}{d}\right) d.$$

cq(n) is multiplicative when considered as a function of q for a fixed value of n:[5] i.e.

$$\text{If } (q,r) = 1 \text{ then } c_q(n)c_r(n) = c_{qr}(n).$$

From the definition (or Kluyver's formula) it is straightforward to prove that, if p is a prime number,

$$c_p(n) = \begin{cases} -1 & \text{if } p \nmid n \\ \phi(p) & \text{if } p \mid n' \end{cases}$$

p^k is a prime power where $k > 1$,

$$c_{p^k}(n) = \begin{cases} 0 & \text{if } p^{k-1} \nmid n \\ -p^{k-1} & \text{if } p^{k-1} \mid n \text{ and } p^k \nmid n. \\ \phi(p^k) & \text{if } p^k \mid n \end{cases}$$

This result and the multiplicative property can be used to prove

$$c_q(n) = \mu\left(\frac{q}{(q,n)}\right)\frac{\phi(q)}{\phi\left(\frac{q}{(q,n)}\right)}.$$

The equivalence of it and Ramanujan's sum is due to Hölder.

Other properties of $c_q(n)$

For all positive integers *q*,

$$\begin{aligned} c_1(q) &= 1 \\ c_q(1) &= \mu(q) \\ c_q(q) &= \phi(q) \\ c_q(m) &= c_q(n) \qquad \text{for } m \equiv n(\text{mod} q) \end{aligned}$$

For a fixed value of *q* the absolute value of the sequence $\{c_q(1), c_q(2), \ldots\}$ is bounded by $\phi(q)$, and for a fixed value of *n* the absolute value of the sequence $\{c_1(n), c_2(n), \ldots\}$ is bounded by *n*.

If $q > 1$

$$\sum_{n=a}^{a+q-1} c_q\,(n) = 0.$$

m_1, $m_2 > 0$, $m = \text{lcm}\,(m_1, m_2)$. Then Ramanujan's sums satisfy an orthogonality property:

$$\frac{1}{m}\sum_{k=1}^{m} c_{m_1}(k)c_{m_2}(k) = \begin{cases} \phi(m) & m_1 = m_2 = m, \\ 0 & \text{otherwise} \end{cases}$$

Let *n*, *k* > 0. Then

$$\sum_{\substack{d|n \\ \gcd(d,k)=1}} d\frac{\mu\left(\frac{n}{d}\right)}{\phi(d)} = \frac{\mu(n)c_n(k)}{\phi(n)},$$

Brauer - Rademacher identity:

If *n* > 0 and *a* is any integer, we also have

$$\sum_{\substack{1 \le k \le n \\ \gcd(k,n)=1}} c_n\,(k-a) = \mu(n)c_n(a),$$

due to Cohen.

4.13 BROCARD–RAMANUJAN DIOPHANTINE EQUATION

Brocard's problem is a problem in mathematics that seeks integer values of n such that $(n! + 1)$ is a perfect square, where $n!$is the factorial. Only three values of n are known: 4, 5, 7 — and it is not known whether there are any more. Though research has extended far beyond n > 7, no additional solutions to the equation $n! + 1 = m^2$ are known.

More formally, it seeks pairs of integers $n$and m, such that

$$n! + 1 = m^2.$$

The problem was posed by Henri Brocard in a pair of articles in 1876 and 1885, and independently in 1913 by Srinivasa Ramanujan.

Pairs of the numbers (n, m) that solve Brocard's problem were named **Brown numbers** by Clifford A. Pickover in his 1995 book, after learning of the problem from Kevin S. Brown. As of October 2022, there are only three known pairs of Brown numbers:

(4,5), (5,11), and (7,71), based on the equalities

$4! + 1 = 5^2 = 25$,

$5! + 1 = 11^2 = 121$, and

$7! + 1 = 71^2 = 5041$.

Paul Erdős conjectured that no other solutions exist. Computational searches up to one quadrillion have found no further solutions.

Connection to the abc-conjecture

It would follow from the abc-conjecture that there are only finitely many Brown numbers. More generally, it would also follow from the abc-conjecture that

$$n! + A = k^2$$

has only finitely many solutions, for any given integer A, and that

$$n! = P(x)$$

has only finitely many integer solutions, for any given polynomial $P(x)$ of degree at least 2 with integer coefficients.

4.14 RAMANUJAN'S PARTITION FUNCTION (NUMBER THEORY)

In one of the letters to Hardy in 1913, Ramanujan gave a formula for the coefficients of a certain series expansion of an infinite product which suggested that there ought to be a similar exact formula for the partition function, p(n), the number of partitions of a positive integer n. Hardy felt that this claim of Ramanujan concerning an exact formula for p(n) in terms of continuous functions was too good to be true but was convinced that it would be possible to construct an asymptotic series expansion. Asymptotic formulas are like approximate formulas. By an ingenious and intricate calculation involving the singularities of the generating function of p(n), Hardy and Ramanujan obtained an asymptotic formula which when calculated up to a certain number of terms yielded a value that differed from p(n) by no more than the fourth root of n. Since p(n) is an integer, it is clear that its value is the nearest integer to what is given by the series. The proof of this formula required sophisticated tools from complex variable theory, and here Hardy's mastery over analytic methods was crucial. The series Hardy and Ramanujan obtained was genuinely an asymptotic series in the sense that when summed to infinity it diverges. Subsequently, Hans Rademacher noticed that by making a very mild but important change, namely, by replacing the exponential functions by hyperbolic functions, the Hardy–Ramanujan asymptotic series could in fact be converted to an infinite series that converges top(n). Actually, in the 1913 letter to Hardy, Ramanujan used hyperbolic functions to claim an exact formula for a related problem. So, Ramanujan was indeed correct in surmising that a similar exact formula would exist for p(n). Professor Atle Selberg of the Institute for Advanced Study in Princeton, one the greatest living mathematicians, said during his talk at the Ramanujan Centennial in Madras in 1987, that the exact formula of Rademacher was indeed more natural than the Hardy–Ramanujan asymptotic formula. Selberg discovered this formula on his own but did not publish this once he found out that Rademacher had obtained it earlier. According to Selberg, even though Ramanujan correctly claimed the existence of an exact formula for p(n), out of respect for his mentor Hardy, he settled for less, namely the asymptotic formula. In any case, the Hardy–Ramanujan asymptotic formula gave rise to a very powerful analytic method to evaluate the coefficients of series arising in a wide class of problems in additive number theory. This circle method, originally due to Hardy–Ramanujan, was subsequently developed by Hardy and Littlewood and others, and is one of the most widely applicable methods today

4.14.1 ON PARTITION FUNCTIONS

In number theory, the partition function p(n) represents the number of possible partitions of a non-negative integer n.

For instance, p(4) = 5 because the integer 4 has the five partitions 1 + 1 + 1 + 1, 1 + 1 + 2, 1 + 3, 2 + 2, and 4.

Example 2: p (5) = 7 for the seven ways of partitioning the number 5:

5, 4 + 1, 3 + 2, 3 + 1 + 1, 2 + 2 + 1, 2 + 1 + 1 + 1, 1 + 1 + 1 + 1 + 1.

No closed-form expression for the partition function is known, but it has both asymptotic expansions that accurately approximate it and recurrence relations by which it can be calculated exactly. It grows as an exponential function of the square root of its argument. The multiplicative inverse of its generating function is the Euler function; by Euler's pentagonal number theorem, this function is an alternating sum of pentagonal number powers of its argument.

Srinivasa Ramanujan first discovered that the partition function has nontrivial patterns in modular arithmetic, now known as Ramanujan's congruences. For instance, whenever the decimal representation of n ends in the digit 4 or 9, the number of partitions of n will be divisible by 5.

Definition and examples

For a positive integer n, p(n) is the number of distinct ways of representing n as a sum of positive integers. For the purposes of this definition, the order of the terms in the sum is irrelevant: two sums with the same terms in a different order (e.g., 1 + 1 + 2 and 1 + 2 + 1) are not considered distinct.

By convention p(0) = 1, as there is one way of representing 0 as a sum of positive integers (the empty sum). Furthermore p(n) = 0 when n is negative.

The first few values of the partition function, starting with p(0) = 1, are:

1, 1, 2, 3, 5, 7, 11, 15, 22, 30, 42, 56, 77, 101, 135, 176, 231, 297, 385, 490, 627, 792, 1002, 1255, 1575, 1958, 2436, 3010, 3718, 4565, 5604, ...

Some exact values of p(n) for larger values of n include:

$$p(100) = 190{,}569{,}292$$

$$\begin{aligned} p(1000) &= 24{,}061{,}467{,}864{,}032{,}622, 473{,}692{,}149{,}727{,}991 \\ &\approx 2.40615 \times 10^{31} \end{aligned}$$

$$\begin{aligned} p(10000) &= 36{,}167{,}251{,}325, \ldots, 906{,}916{,}435{,}144 \\ &\approx 3.61673 \times 10^{106} \end{aligned}$$

4.15 LANDAU–RAMANUJAN CONSTANT

In mathematics and the field of number theory, the **Landau–Ramanujan constant** is the positive real number *b* that occurs in a theorem proved by Edmund Landau in 1908, stating that for large $x$positive integers below x that are the sum of two square numbers behaves asymptotically as:

$$\frac{bx}{\sqrt{\log(x)}}.$$

constant *b* was rediscovered in 1913 by Srinivasa Ramanujan, in the first letter he wrote to G.H. Hardy.

4.15.1 Sums of two squares

By the sum of two squares theorem, the numbers that can be expressed as a sum of two squares of integers are the ones for which each prime number congruent to 3 mod 4 appears with an even exponent in their prime factorization. For instance, *45 = 9 + 36* is a sum of two squares; in its prime factorization, $3^2 \times 5$, the prime 3 appears with an even exponent, and the prime 5 is congruent to 1 mod 4, so its exponent can be odd.

Landau's theorem states that if $N(x)$is the number of positive integers less than x that are the sum of two squares, then

$$\lim_{x\to\infty}\left(\frac{N(x)}{\frac{x}{\sqrt{\log(x)}}}\right) = b \approx$$

$$0.76422365358922066299069873125009232811679054$$

where b is the Landau–Ramanujan constant.

The Landau-Ramanujan constant can also be written as an infinite product:

$$b = \frac{1}{\sqrt{2}} \prod_{p\equiv 3(\mathrm{mod}4)} \left(1 - \frac{1}{p^2}\right)^{-\frac{1}{2}} = \frac{\pi}{4} \prod_{p\equiv 1(\mathrm{mod}4)} \left(1 - \frac{1}{p^2}\right)^{\frac{1}{2}}.$$

$N(x)$ as an integral, with the same constant of proportionality, and with a slowly growing error term.

4.16 HARDY–RAMANUJAN THEOREM

In mathematics, the **Hardy–Ramanujan theorem**, proved by Ramanujan, and checked by Hardy states that the normal order of the number $\omega(n)$of distinct prime factors of a number n is $\log\log n$

Roughly speaking, this means that most numbers have about this number of distinct prime factors.

4.16.1 Precise statement

A more precise version states that for every real-valued function $\psi(n)$that tends to infinity as n tends to infinity

$$|\omega(n) - \log\log n| < \psi(n)\sqrt{\log\log n}$$

$$|\omega(n) - \log\log n| < \ (log \log n)^{\frac{1}{2}+\varepsilon}$$

for almost all (all but an infinitesimal proportion of) integers. That is, let g(x)be the number

of positive integers n less than x for which the above inequality fails: then $\frac{g(x)}{x}$converges to zero as xgoes to infinity.

A simple proof to the result was given by Pál Turán, who used the Turán sieve to prove that

$$\sum_{n\leq x} |\ \omega(n) - \log\log x|^2 \ll \mathrm{x}\log\log x.$$

4.16.2 Generalizations

The same results are true of $\Omega(n)$, the number of prime factors of n counted with multiplicity. This theorem is generalized by the Erdős–Kac theorem, which shows that $\omega(n)$is essentially normally distributed. There are many proofs of this, including the method of moments (Granville & Soundararajan) and Stein's method (Harper). It was shown by Durkan that a modified version of Turán's result allows one to prove the Hardy–Ramanujan Theorem with any even moment.

4.17 RAMANUJAN–SOLDNER CONSTANT

In mathematics, the **Ramanujan–Soldner** constant is a mathematical constant defined as the unique positive zero of the logarithmic integral function. It is named after Srinivasa Ramanujan and Johann Georg von Soldner.

Its value is approximately $\mu \approx$

1.45136923488338105028396848589202744949303228...

Since the logarithmic integral is defined by

$$\mathrm{li}(x) = \int_0^x \frac{dt}{\ln t},$$

$\mathrm{li}(\mu) = 0,$ we have

$$\mathrm{li}(x) = \mathrm{li}(x) - \mathrm{li}(\mu) = \int_0^x \frac{dt}{\ln t} - \int_0^\mu \frac{dt}{\ln t} = \int_\mu^x \frac{dt}{\ln t},$$

μ. Also, since the exponential integral function satisfies the equation $\mathrm{li}(x) = \mathrm{Ei}(\ln x)$,

the only positive zero of the exponential integral occurs at the natural logarithm of the Ramanujan–Soldner constant, whose value is approximately $\ln(\mu) \approx$ 0.372507410781366634 461991866...

4.18 RAMANUJAN'S TERNARY QUADRATIC FORM

In number theory, a branch of mathematics, **Ramanujan's ternary quadratic form** is the algebraic expression $x^2 + y^2 + 10z^2$ with integral values for x, y and z. Srinivasa considered this expression in a footnote in a paper published in 1916 and briefly discussed the representability of integers in this form. After giving necessary and sufficient conditions that an integer cannot be represented in the form $ax^2 + by^2 + cz^2$ for certain specific values of a, b and c, Ramanujan observed in a footnote:

"(These) results may tempt us to suppose that there are similar simple results for the form $ax^2 + by^2 + cz^2$ whatever are the values of a, b and c. It appears, however, that in most cases there are no such simple results."

To substantiate this observation, Ramanujan discussed the form which is now referred to as Ramanujan's ternary quadratic form.

4.18.1 Properties discovered by Ramanujan

In his 1916 paper Ramanujan made the following observations about the form $x^2 + y^2 + 10z^2$.

- The even numbers that are not of the form $x^2 + y^2 + 10z^2$ are $4^{\lambda}(16\mu + 6)$.
- The odd numbers that are not of the form $x^2 + y^2 + 10z^2$, viz. 3, 7, 21, 31, 33, 43, 67, 79, 87, 133, 217, 219, 223, 253, 307, 391, ... do not seem to obey any simple law.

4.18.2 Odd numbers beyond 391

By putting an ellipsis at the end of the list of odd numbers not representable as $x^2 + y^2 + 10z^2$, Ramanujan indicated that his list was incomplete. It was not clear whether Ramanujan intended it to be a finite list or infinite list. This prompted others to look for such odd numbers. In 1927,

Burton W. Jones and Gordon Pall discovered that the number **679** could not be expressed in the form $x^2 + y^2 + 10z^2$ and they also verified that there were no other such numbers below 2000. This led to an early conjecture that the **seventeen numbers** – the sixteen numbers in Ramanujan's list and the number discovered by them – were the only odd numbers not representable as $x^2 + y^2 + 10z^2$. However, in 1941, H Gupta showed that the number 2719 could not be represented as $x^2 + y^2 + 10z^2$. He also verified that there were no other such numbers below 20000. Further progress in this direction took place only after the development of modern computers. W. Galway wrote a computer program to determine odd integers not expressible as $x^2 + y^2 + 10z^2$. Galway verified that there are only eighteen numbers less than 2×10^{10} not representable in the form $x^2 + y^2 + 10z^2$.

Based on Galway's computations, Ken Ono and K. Soundararajan formulated the following conjecture:

The odd positive integers which are not of the form $x^2 + y^2 + 10z^2$ are: 3, 7, 21, 31, 33, 43, 67, 79, 87, 133, 217, 219, 223, 253, 307, 391, 679, 2719.

4.18.3 Some known results

The conjecture of Ken Ono and Soundararajan has not been fully resolved. However, besides the results enunciated by Ramanujan, a few more general results about the form have been established. The proofs of some of them are quite simple while those of the others involve quite complicated concepts and arguments.

- Every integer of the form $10n + 5$ is represented by Ramanujan's ternary quadratic form.
- If n is an odd integer which is not square-free then it can be represented in the form $x^2 + y^2 + 10z^2$.
- There are only a finite number of odd integers which cannot be represented in the form $x^2 + y^2 + 10z^2$.

- If the generalized Riemann hypothesis is true, then the conjecture of Ono and Soundararajan is also true.
- Ramanujan's ternary quadratic form is not regular in the sense of L.E. Dickson.

4.19 RAMANUJAN SUMMATION

Ramanujan summation is a technique invented by the mathematician Srinivasa Ramanujan for assigning a value to divergent infinite series. Although the Ramanujan summation of a divergent series is not a sum in the traditional sense, it has properties that make it mathematically useful in the study of divergent infinite series, for which conventional summation is undefined.

Since there are no properties of an entire sum, the Ramanujan summation functions as a property of partial sums. If we take the Euler–Maclaurin summation formula together with the correction rule using Bernoulli numbers, we see that:[*clarification needed*][*further explanation needed*]

$$\frac{1}{2}f(0)+f(1)+\cdots+f(n-1)+\frac{1}{2}f(n)$$

$$=\frac{f(0)+f(n)}{2}+\sum_{k=1}^{n-1}f(k)=\sum_{k=0}^{n}f(k)-\frac{f(0)+f(n)}{2}$$

$$=\int_0^n f(x)\,dx+\sum_{k=1}^{p}\frac{B_{2k}}{(2k)!}\left[f^{(2k-1)}(n)-f^{(2k-1)}(0)\right]+R_p$$

wrote this again for different limits of the integral and the corresponding summation for the case in which p goes to infinity:

$$\sum_{k=a}^{x}f(k)=C+\int_a^x f(t)dt+\frac{1}{2}f(x)+\sum_{k=1}^{\infty}\frac{B_{2k}}{(2k)!}f^{(2k-1)}(x)$$

C is a constant specific to the series and its analytic continuation and the limits on the integral were not specified by Ramanujan, but presumably they were as given above. Comparing both formulae and assuming that R tends to 0 as x tends to infinity, we see that, in a general case, for functions $f(x)$ with no divergence at $x = 0$:

$$C(a) = \int_0^a f(t)\,dt - \frac{1}{2}f(0) - \sum_{k=1}^{\infty} \frac{B_{2k}}{(2k)!} f^{(2k-1)}(0)$$

$a = 0$. By taking $a = \infty$, we normally recover the usual summation for convergent series.

For functions $f(x)$ with no divergence at $x = 1$, we obtain:

$$C(a) = \int_1^a f(t)\,dt + \frac{1}{2}f(1) - \sum_{k=1}^{\infty} \frac{B_{2k}}{(2k)!} f^{(2k-1)} \qquad (1)$$

alternatively, applying smoothed sums.

The convergent version of summation for functions with appropriate growth condition is then[:

$$f(1) + f(2) + f(3) + \cdots = -\frac{f(0)}{2} + i\int_0^{\infty} \frac{f(it) - f(-it)}{e^{2\pi t} - 1}\,dt$$

Abel–Plana formula.

4.19.1 Ramanujan summation of divergent series

In the following text, $(\Re)$ indicates "Ramanujan summation". This formula originally appeared in one of **Ramanujan's notebooks**, without any notation to indicate that it exemplified a novel method of summation.

For example, the $(\Re)$ of 1 − 1 + 1 − ⋯ is:

$$1 - 1 + 1 - \cdots = \frac{1}{2}(\Re).$$

Ramanujan had calculated "sums" of known divergent series. It is important to mention that the Ramanujan sums are not the sums of the series in the usual sense, i.e. the partial sums do not converge to this value, which is denoted by the symbol ($\Re$). In particular, the ($\Re$), sum of $1+2+3+4+\cdots$ was calculated as:

$$1+2+3+\cdots=-\frac{1}{12}(\Re)$$

$1+2^{2k}+3^{2k}+\cdots=0(\Re)$

and for odd powers the approach suggested a relation with the **Bernoulli numbers:**

$$1+2^{2k-1}+3^{2k-1}+\cdots=-\frac{B_{2k}}{2k}(\Re)$$

$C(1)$ rather than $C(0)$ as the result of Ramanujan's summation, since then it can be assured that one series $\sum_{k=1}^{\infty} f(k)$ admits one and only one Ramanujan's summation, defined as the value in 1 of the only solution of the difference equation $R(x)-R(x+1)=f(x)$that verifies the condition $\int_1^2 R(t)\,dt=0$.

This demonstration of Ramanujan's summation (denoted as $\sum_{n\geq 1}^{\Re} f(n)C(0)$, nor with the summation of convergent series, but it has interesting properties, such as: If $R(x)$ tends to a finite limit when $x \rightarrow 1$, then the series $\sum_{n\geq 1}^{\Re} f(n)$ is convergent, and we have

$$\sum_{n\geq 1}^{\Re} f(n)=\lim_{N\to\infty}\left[\sum_{n=1}^{N} f(n)-\int_1^N f(t)\,dt\right]$$

$$\sum_{n\geq 1}^{\Re}\frac{1}{n}=\gamma$$

where $\gamma$is the Euler–Mascheroni constant.

4.19.2 Extension to integrals

Ramanujan re-summation can be extended to integrals; for example, using the Euler–Maclaurin summation formula, one can write

$$\int_a^\infty x^{m-s}\,dx = \frac{m-s}{2}\,dx + \zeta(s-m) - \sum_{i=1}^{a}[i^{m-s} + a^{m-s}]$$

$$-\sum_{r=1}^{\infty} \frac{B_{2r}\theta(m-s+1)}{(2r)!\,\Gamma(m-2r+2-s)}(m-2r+1-s)\int_a^\infty x^{m-2r-s}\,dx$$

which is the natural extension to integrals of the zeta regularization algorithm.

This recurrence equation is finite, since for $m - 2r < -1$,

$$\int_a^\infty dx\, x^{m-2r} = -\frac{a^{m-2r+1}}{m-2r+1}.$$

Note that this involves

$$I(n,\Lambda) = \int_0^\Lambda dx\, x^n.$$

$\Lambda \to \infty$, the application of this Ramanujan re-summation lends to finite results in the renormalization of quantum field theories.

4.20 RAMANUJAN'S INFINITE SUM

One of the most famous examples of Ramanujan's summation is his handling of the infinite sum of natural numbers:

1 + 2 + 3 + 4 + . . .

Intuitively, this series diverges to infinity.

However, using his summation method, Ramanujan assigned it a surprising and counterintuitive value of −1/12. This result, although not a sum in the conventional sense, is consistent within the framework of Ramanujan's summation and has significant implications in various fields, including theoretical physics and **string theory**.

One way to understand this result is through the regularization of the series using the **Riemann zeta function, ζ(s),** which is defined as:

$$\zeta(s) = \sum_{n=1}^{\infty} \frac{1}{n^s}$$

For Re(s) > 1, this series converges. However, through analytic continuation, it can be extended to other values of sss. For s = −1, we have:

$$\zeta(-1) = -1/12$$

Since the original series 1 + 2 + 3 + 4 + . . . can be formally written as ζ(−1), we get:

$$1 + 2 + 3 + 4 + 5 + \ldots = \zeta(-1)$$

4.21 RAMANUJAN'S MASTER THEOREM

In mathematics, **Ramanujan's master theorem**, named after Srinivasa Ramanujan, is a technique that provides an analytic expression for the **Mellin transform** of an analytic function.

If a complex-valued function $f(x)$has an expansion of the form

$$f(x) = \sum_{k=0}^{\infty} \frac{\varphi(k)}{k!}(-x)^k$$

where $\varphi(s)$is an analytic function, then the Mellin transform of $f(x)$is given by

$$\int_0^{\infty} x^{s-1} f(x)dx = \Gamma(s)\varphi(-s)$$

$\Gamma(s)$ is the gamma function.

It was widely used by Ramanujan to calculate definite integrals and infinite series.

Higher-dimensional versions of this theorem also appear in quantum physics through Feynman diagrams.

A similar result was also obtained by Glaisher.

Alternative formalism

An alternative formulation of Ramanujan's master theorem is as follows:

$$\int_0^\infty x^{s-1}\left(\lambda(0) - x\lambda(1) + x^2\lambda(2) - \cdots\right)dx = \frac{\pi}{\sin(\pi s)}\lambda(-s)$$

$\lambda(n) \equiv \frac{\varphi(n)}{\Gamma(1+n)}$ and using the functional equation for the gamma function.

The integral above is convergent for $0 < \mathcal{R}e(s) < 1$subject to growth conditions on φ.

Proof

A proof subject to "natural" assumptions (though not the weakest necessary conditions) to Ramanujan's master theorem was provided by G. H. Hardy, employing the residue theorem and the well-known Mellin inversion theorem.

Application to Bernoulli polynomials

The generating function of the Bernoulli polynomials $B_k(x)$is given by:

$$\frac{ze^{xz}}{e^z - 1} = \sum_{k=0}^{\infty} B_k(x)\frac{z^k}{k!}$$

These polynomials are given in terms of the Hurwitz zeta function:

$$\zeta(s,a) = \sum_{n=0}^{\infty} \frac{1}{(n+a)^s}$$

$$\zeta(1-n,a) = -\frac{B_n(a)}{n} \qquad \text{for } n \geq 1$$

$$\int_0^{\infty} x^{s-1}\left(\frac{e^{-ax}}{1-e^{-x}} - \frac{1}{x}\right) dx = \Gamma(s)\zeta(s,a)$$

which is valid for $0 < \mathcal{R}e(s) < 1$

Application to the gamma function

$$\Gamma(x) = \frac{e^{-\gamma x}}{x}\prod_{n=1}^{\infty}\left(1+\frac{x}{n}\right)^{-1} e^{\frac{x}{n}}$$

is equivalent to expression

$$\log\,\Gamma(1+x) = -\gamma x + \sum_{k=2}^{\infty}\frac{\zeta(k)}{k}(-x)^k$$

where $\zeta(k)$ is the Riemann zeta function.

Then applying Ramanujan master theorem, we have:

$$\int_0^{\infty} x^{s-1}\frac{\gamma x + \log\Gamma(1+x)}{x^2}\mathrm{d}x = \frac{\pi}{\sin(\pi s)}\frac{\zeta(2-s)}{2-s}$$

$0 < \mathcal{R}e(s) < 1.$

Special cases of

$s = \frac{1}{2}$ and $s = \frac{3}{4}$ are

$$\int_0^{\infty}\frac{\gamma x + \log\Gamma(1+x)}{x^{\frac{5}{2}}}\mathrm{d}x = \frac{2\pi}{3}\zeta\left(\frac{3}{2}\right)$$

$$\int_0^{\infty}\frac{\gamma x + \log\Gamma(1+x)}{x^{\frac{9}{4}}}\mathrm{d}x = \sqrt{2}\frac{4\pi}{5}\zeta\left(\frac{5}{4}\right)$$

Application to Bessel functions

The Bessel function of the first kind has the power series

$$J_\nu(z) = \sum_{k=0}^{\infty} \frac{(-1)^k}{\Gamma(k+\nu+1)k!} \left(\frac{z}{2}\right)^{2k+\nu}$$

$$\frac{2^{\nu-2s}\pi}{\sin(\pi(s-\nu))} \int_0^\infty z^{s-1-\frac{\nu}{2}} J_\nu(\sqrt{z})dz = \Gamma(s)\Gamma(s-\nu)$$

valid for $0 < 2\mathcal{R}e(s) < \mathcal{R}e(\nu) + \frac{3}{2}$

Equivalently, if spherical Bessel function $j_\nu(z)$ is preferred, the formula becomes

$$\frac{2^{\nu-2s}\sqrt{\pi}(1-2s+2\nu)}{\cos(\pi(s-\nu))} \int_0^\infty z^{s-1-\frac{\nu}{2}} j_\nu(\sqrt{z})dz$$
$$= \Gamma(s)\Gamma\left(\frac{1}{2}+s-\nu\right)$$

$0 < 2\mathcal{R}e(s) < \mathcal{R}e(\nu) + 2.$

The solution is remarkable in that it is able to interpolate across the major identities for the gamma function. In particular, the choice of $J_{\frac{1}{2}}(\sqrt{z})$ gives the square of the gamma function, $j_0(\sqrt{z})$ gives the duplication formula, $z^{-\frac{1}{2}}J_1(\sqrt{z})$gives the reflection formula, and fixing to the evaluable $s = \frac{1}{2}$ or $s = 1$gives the gamma function by itself, up to reflection and scaling.

Bracket integration method

The bracket integration method (method of brackets) applies Ramanujan's master theorem to a broad range of integrals. The bracket integration method generates the integrand's series expansion, creates a bracket series, identifies the series coefficient and formula parameters and computes the integral.

Integration formulas

This section identifies the integration formulas for integrand's with and without consecutive integer exponents and for single and double integrals. The integration formula for double integrals may be generalized to any multiple integral. In all cases, there is a parameter value n^*or array of parameter values N^*that solves one or more linear equations derived from the exponent terms of the integrand's series expansion.

Consecutive integer exponents, 1 variable

This is the function series expansion, integral and integration formula for an integral whose integrand's series expansion contains consecutive integer exponents.

$$f(y) = \sum_{n=0}^{\infty} \frac{(-1)^n}{n!} \varphi(n)\, y^n$$

$$\int_0^{\infty} y^{c-1} f(y) dy$$

$$= \int_0^{\infty} \sum_{n=0}^{\infty} \frac{(-1)^n}{n!} \varphi(n)\, y^{n+c-1} dy$$

$$= \Gamma(-n^*)\varphi(n^*)$$

The parameter n^* is a solution to this linear equation.

$$n^* + c = 0, n^* = -c$$

General exponents, 1 variable

Applying the substitution $y = x^a$generates the function series expansion, integral and integration formula for an integral whose integrand's series expansion may not contain consecutive integer exponents.

$$f(x) = \sum_{n=0}^{\infty} \frac{(-1)^n}{n!} \varphi(n)\, x^{an}$$

$$\int_0^{\infty} x^{c-1} f(x) dx$$

$$= \int_0^{\infty} \sum_{n=0}^{\infty} \frac{(-1)^n}{n!} \varphi(n)\, x^{an+c-1} dx$$

$$= a^{-1}\, \Gamma(-n^*) \varphi(n^*).$$

The parameter $n^*$is a solution to this linear equation.

$$a\, n^* + c = 0, n^* = -a^{-1} c$$

Consecutive integer exponents, double integral

This is the function series expansion, integral and integration formula for a double integral whose integrand's series expansion contains consecutive integer exponents.

$$f(y_1, y_2) = \sum_{n=0}^{\infty} \frac{(-1)^{n_1}}{n_1!} \frac{(-1)^{n_2}}{n_2!} \varphi(n_1, n_2)\, y_1^{n_1}\, y_2^{n_2}$$

$$\int_0^{\infty} y_1^{c_1-1}\, y_2^{c_2-1}\, f(y_1, y_2)\, dy_1\, dy_2$$

$$= \int_0^{\infty} \int_0^{\infty} \sum_{n_1=0}^{\infty} \sum_{n_2=0}^{\infty} \frac{(-1)^{n_1}}{n_1!} \frac{(-1)^{n_2}}{n_2!} \varphi(n_1, n_2)\, y_1^{n_1+c_1-1}$$
$$y_2^{n_2+c_2-1}\, dy_1\, dy_2$$

$$= \Gamma(-n_1^*)\, \Gamma(-n_2^*)\, \varphi(n_1^*, n_2^*)$$

The parameters n_1^* and $n_2^*$are solutions to these linear equations.

$$n_1^* + c_1 = 0, n_2^* + c_2 = 0, n_1^* = -c_1, n_2^* = -c_2$$

General exponents, double integral

This section describes the integration formula for a double integral whose integrand's series expansion may not contain consecutive integer exponents. Matrices contain the parameters needed to express the exponents in a series expansion of the integrand, and the determinant of invertible matrix A is $\det|A|$

$$A = \begin{vmatrix} a_{11} & a_{12} \\ a_{21} & a_{22} \end{vmatrix}, C = \begin{vmatrix} c_1 \\ c_2 \end{vmatrix}, N^* = \begin{vmatrix} n_1^* \\ n_2^* \end{vmatrix}$$

Applying the substitution

$$y_1 = x_1^{a_{11}} x_2^{a_{21}}, y_2 = x_1^{a_{12}} x_2^{a_{22}}$$

generates the function series expansion, integral and integration formula for a double integral whose integrand's series expansion may not contain consecutive integer exponents. The integral and integration formula are

$$\int_0^\infty \int_0^\infty \sum_{n_1=0}^\infty \sum_{n_2=0}^\infty \frac{(-1)^{n_1}}{n_1!} \frac{(-1)^{n_2}}{n_2!} \varphi(n_1, n_2)$$
$$x_1^{n_1 a_{11} + n_2 a_{12} + c_1 - 1} x_2^{n_1 a_{21} + n_2 a_{22} + c_2 - 1} \, dx_1 \, dx_2$$
$$= \det |A|^{-1} \, \Gamma(-n_1^*) \, \Gamma(-n_2^*) \, \varphi(n_1^*, n_2^*).$$

The parameter matrix N^* is a solution to this linear equation.

$$AN^* + C = 0, N^* = -A^{-1}C$$

- *number of brackets* is the number of linear equations associated with an integral. This term reflects the common practice of bracketing each linear equation.
- The *complexity index* is the number of integrand sums minus the number of brackets (linear equations). Each series expansion of the integrand contributes one sum.
- The *summation indices (variables)* are the indices that index terms in a series expansion. In the example, there are 3

summation indices n_1, n_2 and $n_3$because the integrand is a product of 3 series expansions.

- The *free summation indices (variables)* are the summation indices that remain after completing all integrations. Integration reduces the number of sums in the integrand by replacing the series expansions (sums) with an integration formula. Therefore, there are fewer summation indices after integration. The number of chosen free summation indices equals the complexity index.

Integrals with a positive complexity index

The free summation indices $\overline{n}_1, \dots, \overline{n}_f$ are elements of set $F\overline{N}$ and the coefficients of the free summation indices is matrix $\overline{A}$

$$\overline{A} = \begin{vmatrix} \overline{a}_{11} & \dots & \overline{a}_{1f} \\ \vdots & & \vdots \\ \overline{a}_{b1} & \dots & \overline{a}_{bf} \end{vmatrix}, \overline{N} = \begin{vmatrix} \overline{n}_1 \\ \vdots \\ \overline{n}_f \end{vmatrix}$$

The remaining indices are set Bcontaining indices $n_1, \dots, n_b$. Matrices $A, C$and N^* contain matrix elements that multiply or sum with the non-summation indices. The selected free summation indices must leave matrix Anon-singular.

$$A = \begin{vmatrix} a_{11} & \dots & a_{1b} \\ \vdots & & \vdots \\ a_{b1} & \dots & a_{bb} \end{vmatrix}, C = \begin{vmatrix} c_1 \\ \vdots \\ c_b \end{vmatrix}, N^* = \begin{vmatrix} n_1^* \\ \vdots \\ n_b^* \end{vmatrix}$$

This is the function's series expansion, integral and integration formula.

$$f(x_1, \dots, x_b)$$

$$= \sum_{n \in B}^{\infty} \sum_{\overline{n} \in F}^{\infty} \frac{(-1)^{n_1}}{n_1!} \frac{(-1)^{\overline{n}_1}}{\overline{n}_1!} \dots \frac{(-1)^{n_b}}{n_b!} \frac{(-1)^{\overline{n}_f}}{\overline{n}_f!} \varphi(n_1, \dots, n_b, \overline{n}_1, \dots, \overline{n}_f) \prod_{x_k \in B} x_k^{n_1 a_{k1} + \dots + \overline{n}_1 \overline{a}_{k1} + \dots + c_k - 1}$$

$$\int_0^{\infty} \dots \int_0^{\infty} x^{c_1 - 1} \dots x^{c_b - 1} f(x_1, \dots, x_b)\, dx_1 \dots dx_b$$

$$= \det |A|^{-1} \sum_{\overline{n} \in F}^{\infty} \frac{(-1)^{\overline{n}_1}}{\overline{n}_1!} \dots \frac{(-1)^{\overline{n}_f}}{\overline{n}_f!} \Gamma(-n_1^*) \dots \Gamma(-n_b^*)\, \varphi(n_1^*, \dots, n_b^*, \overline{n}_1, \dots, \overline{n}_f).$$

The parameters $n_1^*, \dots, n_b^*$ are linear functions of the parameters $\overline{n}_1^*, \dots, \overline{n}_f^*$

$$A\,N^* + \overline{A}\,\overline{N} + C = 0, N^* = -A^{-1}(\overline{A}\,\overline{N} + C).$$

Bracket series

Table 1. Bracket series notations

Notation type	Power series notation	Bracket series notation
Indicator	$\frac{(-1)^n}{n!}$	ϕ_n
Multi-indicator	$\prod_{j=1}^{N}\left(\frac{(-1)^{n_j}}{n_j!}\right)$	$\phi_{n_1,\dots,n_N}$
Bracket	$\int_0^\infty dx\ x^{a_1n_1+\cdots+a_mn_m+c-1}$	$\langle a_1n_1+\cdots+a_mn_m+c\rangle$

Bracket series notations are notations that substitute for common power series notations (Table 1). Replacing power series notations with bracket series notations transforms the power series to a bracket series. A bracket series facilitates identifying the formula parameters needed for integration. It is also recommended to replace a sum raised to a power:

$$\frac{1}{(x_1+\cdots+x_b)^\alpha}$$

with this bracket series expression:

$$\sum_{m_1=0}^{\infty}\cdots\sum_{m_b=0}^{\infty}\phi_{m_1,\dots,m_b}\,x_1^{m_1}\dots x_b^{m_b}\,\frac{\langle\alpha+m_1+\cdots+m_b\rangle}{\Gamma(\alpha)}.$$

Algorithm

This algorithm describes how to apply the integral formulas.

Table 2. Integral formulas

Complexity index	**Integral formula**
Zero, single integral	$a^{-1}\,\Gamma(-n^*)\varphi(n^*)$
	$\det\,\lvert A\rvert^{-1}\,\Gamma(-n_1^*)\ldots\Gamma(-n_b^*)\;\varphi(n_1^*,\ldots,n_b^*)$
Positive	$\det\,\lvert A\rvert^{-1}\sum_{\bar{n}\in F}^{\infty}\frac{(-1)^{\bar{n}_1}}{\bar{n}_1!}\cdots\frac{(-1)^{\bar{n}_f}}{\bar{n}_f!}\,\Gamma(-n_1^*)\ldots\Gamma(-n_b^*)$ $\varphi(n_1^*,\ldots,n_b^*,\bar{n}_1,\ldots,\bar{n}_f)$

Integral expression

Output Integral value or integral cannot be assigned a value

1. Express the integrand as a power series.
2. Transform the integrand's power series to a bracket series.
3. Obtain the complexity index, formula parameters and series coefficient function.

Complexity index is the number of integrand sums minus number of brackets.

1. Parameters n^* or array $N^*$are solutions to linear equations $an^* + c = 0$
2. (zero complexity index, single integral), $AN^* + C = 0$ (zero complexity index, single integral) or $AN^* + \overline{AN} + C = 0$ (positive complexity index).
3. Identify parameter $a$or (zero complexity index, single integral) or compute $\det|A|$

(all other cases) from the associated linear equations. Identify the series coefficient function $\varphi(n)$of the bracket series.

1. If the complexity index is negative, return integral cannot be assigned a value.
2. If the complexity index is zero, select the formula from table 2 for zero complexity index, single or multiple integral, compute the integral value with this formula, and return this integral value.
3. If the complexity index is positive, select the formula from table 2 for positive complexity index, and compute the integral value as a series expansion with this formula for all possible choices of the free summation indices. Select the lowest complexity index, convergent series expansion, adding series that converge in the same region.
    1. If all series expansions are divergent series or null series (all series terms zero), then return integral cannot be assigned a value.
    2. If the series expansion is non-null and non-divergent, return this series expansion as the integral value.

Examples

Zero complexity index

The bracket method will integrate this integral.

$$\int_0^\infty x^{\frac{3}{2}}\, e^{-\frac{x^3}{2}}\, dx$$

Express the integrand as a power series.

$$\int_0^\infty \sum_{n=0}^{\infty} 2^{-n}\, \frac{(-1)^n}{n!}\, x^{\left(3\cdot n+\frac{5}{2}\right)-1}\, dx$$

$$\sum_{n=0}^{\infty} 2^{-n}\, \phi(n) \cdot \left\langle 3\, n + \frac{5}{2} \right\rangle$$

Obtain the complexity index, formula parameters and series coefficient function.

Complexity index is zero.

$$3\,n^* + \frac{5}{2} = 0$$

$n^* = -\frac{5}{6}, a = 3$

$$\varphi(n) = 2^{-n}$$

$$\int_0^\infty x^{\frac{3}{2}} \cdot e^{-\frac{x^3}{2}}\,dx$$

$$= a^{-1}\,\Gamma(-n^*)\,\varphi(n^*)$$

$$= \frac{\Gamma\left(\frac{5}{6}\right)\,2^{\frac{5}{6}}}{3}$$

Positive complexity index

The bracket method will integrate this integral.

$$\int_0^\infty \frac{1}{\left(1+x^3+x^5\right)^{\frac{1}{2}}}\,dx$$

1. Express the integrand as a power series. Use the sum raised to a power formula.

$$\int_0^\infty \sum_{n_1,n_2,n_3} \frac{1}{\sqrt{\Gamma\left(\frac{1}{2}\right)}} \phi_{123} 1^{n_1} x^{5n_2+3n_3} \left\langle n_1 + n_2 + n_3 + \frac{1}{2}\right\rangle dx$$

2. Transform the power series to a bracket series.

$$\int_0^\infty \sum_{n_1,n_n,n_3} \frac{1}{\sqrt{\Gamma\left(\frac{1}{2}\right)}} \phi_{123} \langle 5\,n_2 + 3\,n_3 + 1\rangle \left\langle n_1 + n_2 + n_3 + \frac{1}{2}\right\rangle$$

3. Obtain the complexity index, formula parameters and series coefficient function.

Complexity index is 1 as 3 sums and 2 brackets.

Select n_3 as the free index, $\overline{n}_3$

$$5n_2^* + 3\overline{n}_3 + 1 = 0, n_1^* + n_2^* + \overline{n}_3 + \frac{1}{2} = 0$$

$$\begin{vmatrix} 1 & 1 \\ 0 & 5 \end{vmatrix} \begin{vmatrix} n_1^* \\ n_2^* \end{vmatrix} + \begin{vmatrix} 1 \\ 3 \end{vmatrix} |\overline{n}_3| + \begin{vmatrix} \frac{1}{2} \\ 1 \end{vmatrix} = 0$$

$$AN^* + \overline{AN} + C = 0$$

$$\det|A| = 5$$

$$n_1^* = -\frac{2}{5}\overline{n}_3 - \frac{3}{10}, n_2^* = -\frac{3}{5}\overline{n}_3 - \frac{1}{5}.$$

$$\varphi(n_1^*, n_2^*, \overline{n}_3) = \frac{1}{\sqrt{\Gamma\left(\frac{1}{2}\right)}} = \frac{1}{\sqrt{\pi}}$$

4. Use table 2 to compute the integral

$$\int_0^\infty \frac{1}{\left(1+x^3+x^5\right)^{\frac{1}{2}}} dx$$

$$= \sum_{\overline{n}_3=0}^{\infty} \frac{(-1)^{\overline{n}_3}}{\overline{n}_3!} \det |A|^{-1} \Gamma(-n_1^*)\Gamma(-n_2^*)\varphi(n_1^*, n_2^*, \overline{n}_3)$$

$$= \sum_{\overline{n}_3=0}^{\infty} \frac{(-1)^{\overline{n}_3}}{\overline{n}_3!} \frac{\Gamma\left(\frac{2}{5}\overline{n}_3 + \frac{3}{10}\right)\Gamma\left(\frac{3}{5}\overline{n}_3 + \frac{1}{5}\right)}{5\sqrt{\pi}}$$

4.22 RAMANUJAN'S ELLIPTIC INTEGRALS

Srinivasa Ramanujan made profound contributions to the theory of elliptic integrals, particularly through his work on modular equations, the theory of theta functions, and hypergeometric series. He developed innovative inversion formulas for elliptic integrals of signatures 2, 3, 4, and 6, and discovered numerous modular identities connecting incomplete elliptic integrals of the first kind to Rogers–Ramanujan continued fractions.

4.22.1 Key aspects of Ramanujan's work on elliptic integrals include:

- **Inversion Formulas:** Ramanujan found formulas to invert elliptic integrals (express the modulus in terms of the integral) for specific signatures, which are crucial for finding modular transformations, as shown in studies of his notebook entries.
- **Modular Relations:** He utilized parameters such as (where is the Rogers–Ramanujan continued fraction) to prove identities for incomplete elliptic integrals of the first kind, especially in the lost notebook.
- **Fastest Series for $\frac{1}{\pi}$:** Ramanujan's studies on complete elliptic integrals E(k) and K(k) led to his famous rapidly converging series for $\frac{1}{\pi}$, which are closely linked to modular equations associated with the $\Gamma 0(N)$, the group gamma function.
- **Generalised Integrals:** He introduced alternative elliptic integrals that produce modular functions, which are currently being re-evaluated for their connections to Ramanujan-type series.

4.23 RAMANUJAN'S HIGHLY COMPOSITE NUMBERS

Ramanujan's Highly Composite Numbers (HCNs) are positive integers that have more divisors than any smaller positive integer (for all). Ramanujan studied these in a 1915 paper, finding over 100 such numbers, including 2, 4, 6, 12, 24, 36, 48, 60, 120, 180, 240, 360, and 720.

4.23.1 Key Aspects of Ramanujan's Highly Composite Numbers:

Definition: A number N is highly composite if M< N implies d(M) < d(N), where d (n) is the number of divisors.

Structure: They are formed by products of prime powers, generally following the pattern: $2^{a_2} . 3^{a_3} . 5^{a_5}$. ,where the exponents are non-increasing.

Significance: These numbers are foundational for studying the maximum order of the divisor function, bridging number theory with the prime number theorem and the Riemann hypothesis.

Examples: The first few highly composite numbers are 1, 2, 4, 6, 12, 24, 36, 48, 60, 120, 180, 240, 360, 720, 840, 1260, 1680, 2520, 5040.

Superior Highly Composite Numbers: Ramanujan also identified a subset called "superior highly composite numbers," which are defined by optimizing the ratio of divisors to a power of the number itself.

Connection to Other Concepts: Every highly composite number is a practical number, meaning all smaller integers can be formed by summing distinct divisors of the number.

Note on Confusion with "Ramanujan Number": It is important to distinguish Ramanujan's *Highly Composite Numbers* from the "Ramanujan number" or "Hardy-Ramanujan number 1729," which is the smallest number expressible as the sum of two cubes in two different ways.

4.24 RAMANUJAN AND π

The number π, the ratio of the circumference of a circle to its diameter, has fascinated mathematicians through the centuries and continues to intrigue researchers even today. Giants like Aryabhata, Archimedes, Newton, Euler and Gauss have through their seminal work contributed substantially to our current understanding of the properties of π and shown how this has a direct bearing on many fundamental questions. And to the impressive list of luminaries who have studied this number, Ramanujan's name must be added. As is usually the case when Ramanujan confronts a topic, he provides a touch of magic to it by means of his incredibly beautiful formulae. In a famous paper published in 1914 in *The Quarterly Journal of Mathematics (Oxford)* entitled "Modular equations and approximations to π", Ramanujan has several tantalising formulae involving π and other numbers, and such expressions are the basis for recent computations of the digits of π **to over two billion decimal places!**

This paper contains work done by Ramanujan in India prior to his departure to England. It is amazing that Ramanujan, who in rural India wrote many of these formulae on a piece of slate and erased them with his elbow, should remain alive in the modern world of the computer! In this article

Early History The realisation that the ratio of the circumference of a circle to its diameter is the same for all circles, is an important landmark in human history. This invariant number is denoted by the Greek letter π (pi). One finds in the Egyptian Rhind Papyrus, which dates about 2000 BC, the approximate value (16/9)2 = 3.1604 ... for π. With regard to the creation of the earth, implicit in The Bible is the statement that π is nearly equal to 3. In attempting to compute the circumference of the earth, the ancients were motivated to calculate π to a high degree of accuracy. Eratosthenes of Alexandria, who is remembered mainly for the Sieve, a procedure to generate prime numbers, actually computed the circumference of the earth. But the one figure from that era who towers above everyone is Archimedes of Syracuse (287–212 BC). Indeed, Archimedes is considered to be one of the five greatest scientific thinkers of all time along with Newton, Euler, Gauss and Einstein. Archimedes was a master of approximation and of the limit process, and in the course of computing the areas and volumes of various geometrical figures, he even anticipated Newton and Leibniz in the development of integral calculus.

By computing the lengths of the inscribed and circumscribed polygon of 96 sides for a circle of unit radius, Archimedes showed that π was less than 31/ 7 and greater than 310/ 71 . We realise today that these early calculations of Archimedes are indeed the first few steps in the harmonic–geometric mean iteration which can be programmed in the computer to give remarkably good approximations to π. Squaring the Circle The members of the Pythagorean school believed that all phenomena could be expressed in terms of integers (whole numbers) and rationals (ratios of integers). For those steeped in this philosophy, it must have come as shock when the square root of 2 was shown to be irrational (not rational). The number π also resisted all attempts to produce an exact rational value, increasing the suspicion that it too might be an irrational number. But the irrationality of π was proved only in 1761. Struggling to understand π geometrically, the Greeks posed the problem of squaring the circle. More precisely, the problem was to construct using only the ruler and compass, a square which is equal in area to a given circle. Since the area of a circle of unit radius is π, what was required was the construction of the side of the corresponding square which will be the square root of π units in length. This is one of the three problems of antiquity. The second problem is to trisect any given angle using only the ruler and compass. (It is easy to bisect any given angle, and indeed this construction is taught in the early years of high school.) The third problem is to double the cube, that is construct a cube, again with only ruler and compass, which is twice the volume of a given cube. This is equivalent to asking for the construction of the cube root of 2 using ruler and compass. The geometrical construction of the square root of 2 is easy, because this is the hypotenuse of a right-angled isosceles triangle whose equal sides are of unit length, a fact known to any high school student of Euclidean Geometry. All three problems of antiquity are now known to be impossible because of the pioneering work of Galois in the 19th century, whose study of the solutions of algebraic equations laid the foundations of Group Theory.

From the work of Gallio, it followed that the only numbers which could be constructed using ruler and compass are special types of algebraic numbers. (Algebraic numbers are those which arise as solutions of polynomial equations with integer coefficients.) Both the square root and the cube root of 2 are algebraic numbers, but the cube root of 2 is not of the special type. In the case of the problem of squaring the circle, its impossibility is a consequence of the fact that π is not an 16 Ramanujan and π 105 algebraic number. This was proved by Lindemann in 1882 and is considered to be one of the crowning achievements of 19th century. Ramanujan was very much interested in the problem of squaring the circle.

In a note published in the *Journal of the Indian Mathematical Society (1913),* Ramanujan offered a geometrical construction to obtain an approximation for the square root of π based on the observation that π is nearly **355/113**. Ramanujan discusses this approximation again in his famous paper of 1914. The most commonly used rational approximation to π is 22/7, and we understand this now in terms of the continued fraction for π. The number 3 is the first approximation that emerges from the continued fraction, with 22/7 as the second (and better) approximation. The number 355/113 considered by Ramanujan is the third approximation. Each successive approximation from the continued fraction is better than the preceding one. Even before the 15th century, the Chinese and Indian mathematicians were aware that 355/113 was an extremely good approximation to π. Representations for π The invention of Calculus paved the way for our present understanding of π and other numbers.

Between 1665 and 1666, Newton himself calculated π to about 15 decimal places by means of an infinite series for the arcsine function. His contemporary and rival in mainland Europe, Leibniz, produced in 1674 a more elegant expression for π using the inverse of the tangent function, a fact that was observed independently by the Scottish mathematician James Gregory in 1671. Indeed, the Gregory series formed the basis for the calculation of π to 71 decimal places by the British astronomer Edmund Halley and his student Abraham Sharp. Other infinite expressions for π were provided during this period. One of the most beautiful and well-known expressions is an infinite product due to John Wallis involving the even numbers in the numerator and the odd numbers in the denominator. In fact, Wallis challenged Lord Brouncker by saying "I bet you can't top this!" Lord Brouncker, who was the first president of The Royal Society, was not a mathematician. Nevertheless, he accepted a challenge and produced a lovely continued fraction expansion for π. Since Lord Brouncker did not give a proof of his derivation, it remained a mystery as to how he arrived at his result. Bruce Berndt points out that Ramanujan has several fascinating continued fractions in his notebooks. One of these continued fraction formulas of Ramanujan for a ratio of gamma functions yields Lord Brouncker's fraction as a special case by setting the variable $x = 1$. Interestingly, Ramanujan had communicated this continued fraction along with many other results in his first letter to Hardy in 1913. Although these infinite expressions for π due to Wallis and his contemporaries were important in understanding many fundamental problems, no one at that time was able to use these formulae to prove that **π is irrational**. Leonard Euler (1707– 1783), the most prolific mathematician in history, was the supreme master in the manipulation of infinite expressions. He produced what is perhaps considered to be the most beautiful and important formula in all of mathematics connecting e, the natural base of the logarithms, i, the imaginary square root of minus 1, and π. Euler's formula is that e to the power iπ equals minus 1.

Using his superior powers of calculation, Euler evaluated several infinite series and products in terms of π. The first proof of the irrationality of π was supplied by Lambert in 1761. Legendre subsequently improved on this and showed that the square of π is irrational, and expressed the opinion that π may not even be an algebraic number, a belief that was shared by Euler. Transcendental numbers, that is numbers which are not algebraic, were not even known to exist at that time. The first transcendental numbers were constructed by Liouville only in 1840. Then in 1873, Charles Hermite showed that e, the natural base of the logarithms, is transcendental. Finally, in 1882, Lindemann, extending the ideas of Hermite, and using Euler's formula, established that π is a transcendental number and thus settled the 2300 year old problem of squaring the circle.

4.24.1 Ramanujan Equation for evaluation of π

Ramanujan's most famous 1910 formula for **π** is a rapidly converging infinite series that adds roughly eight correct decimal places per term.

Published in 1914, the formula is:

$$1/\pi=(2\sqrt{2})/\ 9801 \sum_{n=0}^{\infty} (4n)![(1103+26390n)/[(n!)^4 396^{4n}.]$$

Key Features of this Equation:

- **Rapid Convergence:** Each term in the summation increases the precision by 8 decimal places.
- **Historical Significance:** Ramanujan, an Indian mathematician, developed this and 16 other series for **π**
- **Modern Use:** This formula (specifically the Chudnovsky algorithm derived from it) is used by computers to calculate digits of **π**

SECTION-3

APPLICATIONS OF RAMANAJUN'S FINDINGS IN MODERN SCIENCE & TECHNOLOGY

CHAPTER 5

Impact of Ramanujan's findings in Modern Science & Technology

5.1 RAMANUJAN'S LEGACY IN THE AGE OF AI

Srinivasa Ramanujan, an eminent mathematician, emerged from modest origins to attain a revered status in the field of mathematics during the twentieth century.

He was born in 1887 in Erode, India, and is renowned for his exceptional commitment, innate aptitude, and profound passion for mathematics.

Despite his limited formal education and financial challenges, Ramanujan's remarkable aptitude for mathematics was clear, setting him apart in his area.

He developed thousands of results independently, many of which were later proven to be novel and profoundly significant.

Ramanujan's mathematics was not just a collection of theorems and formulas; it was an art form, brimming with creativity and intuition.

He worked extensively in fields such as infinite series, number theory, continued fractions, q-series, and modular functions.

One of his most famous contributions is the infinite series for π (pi), which not only displayed his mathematical ingenuity but also revolutionised the way calculations for π were done, contributing significantly to the field.

His notebooks, filled with theorems, conjectures, and equations, were decades ahead of their time. For instance, his work on mock theta functions, initially a mystery to contemporary mathematicians, found applications in areas as diverse as number theory, representation theory, and even quantum mechanics.

The comparison of Ramanujan's intuitive and self-taught approach to mathematics with the structured and data-driven processes of modern artificial intelligence (AI) offers a fascinating study of contrasts.

While AI, with its vast computational capabilities and algorithmic precision, represents a pinnacle of technological progress, Ramanujan's story is a testament to the untapped potential of the human mind and its ability to transcend the limitations of its environment through sheer intellect and passion.

This article examines the similarities and differences between the cognitive processes of AI, a creature of structured logic and enormous data processing capability, and Ramanujan, someone whose life was a combination of struggle and brilliance.

We take a closer look at how each has made a distinct contribution to the fields of mathematics and technology, influencing our understanding in ways that were previously unthinkable.

Intuitive Insight Vs Data-Driven Logic

The comparison between Ramanujan's intuitive mathematical insights and AI's data-driven logic illustrates two distinct approaches to mathematics.

Ramanujan, largely self-taught, relied on intuition and creativity, making significant contributions to number theory and infinite series for π (pi) through intuitive leaps rather than systematic analysis.

In contrast, AI employs data-driven logic and algorithms, excelling in pattern recognition and problem-solving within large datasets, such as optimising logistical routes.

The interplay of Ramanujan's intuition and AI's computation highlights the potential for combining these methods for groundbreaking discoveries.

While Ramanujan's approach was characterised by personal insight and serendipity, AI's strength lies in processing data and executing algorithms efficiently.

Together, they represent a fascinating dialogue between human creativity and technological progress, each contributing uniquely to the advancement of mathematics and technology.

Self-Taught Learning Vs Structured AI Algorithms

The difference between Ramanujan's intuitive, self-learn approach to mathematics and the structured, data-centric learning of AI highlights two different ways of acquiring knowledge.

Ramanujan, starting with only basic textbooks, explored deeper into mathematics, driven by his natural curiosity and talent. He made groundbreaking discoveries, such as the Rogers-Ramanujan identities, without formal guidance.

In contrast, AI relies on structured algorithms and data, learning from patterns in large datasets to create predictive models and optimise solutions, yet lacking Ramanujan's ability for intuitive leaps.

This juxtaposition highlights the unique strengths and limitations of human and machine intelligence. Ramanujan's exploratory approach shows the power of human curiosity and innovation, while AI excels in processing data and algorithmic precision.

The potential synergy between the two, like using AI to explore Ramanujan's unproven conjectures, illustrates how they can complement each other in advancing knowledge and technology.

Originality And Creativity: The Human Advantage

The contrast between Ramanujan's creative, original approach to mathematics and the capabilities of AI highlights the divide between human and machine intelligence.

Ramanujan's work, such as his innovative series for π and the partition function of integers, demonstrated a level of originality and creative problem-solving beyond the current scope of AI.

AI excels in pattern recognition and data optimisation but lacks the human capacity for true innovation and the creation of novel concepts.

While AI operates within its programmed parameters and can't emulate human creativity, Ramanujan's legacy inspires the development of new AI algorithms. His mathematical innovations encourage AI applications in unexplored areas and complex problem-solving.

Ultimately, Ramanujan's work underscores the unique human ability for creative thought, highlighting the complementary roles of human ingenuity and AI in advancing various fields.

Ethical And Philosophical Aspects: Beyond Computation

Exploring the differences between Ramanujan's mathematical approach and AI reveals a clear division between human and machine intelligence, especially in ethical and philosophical areas.

Ramanujan's work was deeply personal and philosophical, often linked to spiritual inspiration and seeing mathematics as an art. In stark differentiation, AI, while technically advanced, lacks the capacity for ethical understanding and philosophical reflection, highlighting a distinct nature of intelligence.

AI operates on algorithms and data without the ability to make ethical judgements or engage with philosophical concepts, a limitation evident in areas requiring moral and ethical decision-making, such as healthcare or judicial sentencing.

In contrast, Ramanujan's mathematical journey demonstrates the profound capacity of human intelligence to intertwine computation with ethical and philosophical understanding, underscoring the unique value of human insight in areas beyond the scope of AI.

5.2 RAMANUJAN MACHINE

The **Ramanujan machine** is a specialised software package, developed by a team of scientists at the Technion: Israeli Institute of Technology, to discover new formulas in mathematics. It has been named after the Indian mathematician Srinivasa Ramanujan because it supposedly imitates the thought process of Ramanujan in his discovery of hundreds of formulas. The machine has produced several conjectures in the form of continued fraction expansions of expressions involving some of the most important constants in mathematics like e and π (pi). Some of these conjectures produced by the Ramanujan machine have subsequently been proved true. The others continue to remain as conjectures. The software was conceptualised and developed by a group of undergraduates of the Technion under the guidance of Ido Kaminer, an electrical engineering faculty member of Technion. The details of the machine were published online on 3 February 2021 in the journal *Nature*.

According to George Andrews, an expert on the mathematics of Ramanujan, even though some of the results produced by the Ramanujan machine are amazing and difficult to prove, the results produced by the machine are not of the calibre of Ramanujan and so calling the software the *Ramanujan machine* is slightly outrageous. Doron Zeilberger, an Israeli mathematician, has opined that the Ramanujan machine is a harbinger of a new methodology of doing mathematics.

Using AI and computer automation, Technion researchers have developed a "conjecture generator" that creates mathematical conjectures, which are considered to be the starting point for developing mathematical theorems. They have already used it to generate a number of previously unknown formulas. The study, which was published in the journal Nature, was carried out by undergraduates from different faculties under the tutelage of Assistant Professor Ido Kaminer of the Andrew and Erna Viterbi Faculty of Electrical Engineering at the Technion.

The project deals with one of the most fundamental elements of mathematics – mathematical constants. A mathematical constant is a number with a fixed value that emerges naturally from different mathematical calculations and mathematical structures in different fields. Many mathematical constants are of great importance in mathematics, but also in disciplines that are external to mathematics, including biology, physics, and ecology. The golden ratio and Euler's number are examples of such fundamental constants.

Perhaps the most famous constant is pi, which was studied in ancient times in the context of the circumference of a circle. Today, pi appears in numerous formulas in all branches of science, with many math aficionados competing over who can recall more digits after the decimal point: 3.14159265358979323846264338327950288419716939937510582097494.

The Technion researchers proposed and examined a new idea: The use of computer algorithms to automatically generate mathematical conjectures that appear in the form of formulas for mathematical constants.

A conjecture is a mathematical conclusion or proposition that has not been proved; once the conjecture is proved, it becomes a theorem. Discovery of a mathematical conjecture on fundamental constants is relatively rare, and its source often lies in mathematical genius and exceptional human intuition. Newton, Riemann, Goldbach, Gauss, Euler, and Ramanujan are examples of such genius, and the new approach presented in the paper is named after Srinivasa Ramanujan.

One of Ramanujan's rare capabilities was the intuitive formulation of unproven mathematical formulas. The Technion research team, therefore, decided to name their algorithm "the Ramanujan Machine," as it generates conjectures without proving them, by "imitating" intuition using AI and considerable computer automation.

According to Prof. Kaminer, "Our results are impressive because the computer doesn't care if proving the formula is easy or difficult, and doesn't base the new results on any prior mathematical knowledge, but only on the numbers in mathematical constants. To a large degree, our algorithms work in the same way as Ramanujan himself, who presented results without proof. It's important to point out that the algorithm itself is incapable of proving the conjectures it found – at this point, the task is left to be resolved by human mathematicians."

The conjectures generated by the Technion's Ramanujan Machine have delivered new formulas for well-known mathematical constants such as pi, Euler's number (e), (which is related to the Riemann zeta function), and the Catalan constant. Surprisingly, the algorithms developed by the Technion researchers succeeded not only in creating known formulas for these famous constants, but in discovering several conjectures that were heretofore unknown. The researchers estimate this algorithm will be able to significantly expedite the generation of mathematical conjectures on fundamental constants and help to identify new relationships between these constants.

According to the researchers, "Similar ideas can in the future lead to the development of mathematical conjectures in all areas of mathematics, and in this way provide a meaningful tool for mathematical research."

The research team has launched a website, Ramanujan Machine.com, which is intended to inspire the public to be more involved in the advancement of mathematical research by providing algorithmic tools that will be available to mathematicians and the public at large. Even before the article was published, hundreds of students, experts, and amateur mathematicians had signed up to the website.

The research study started out as an undergraduate project in the Rothschild Scholars Technion Program for Excellence with the participation of Gal Raayoni and George Pisha, and continued as part of the research projects conducted in the Andrew and Erna Viterbi Faculty of Electrical Engineering with the participation of Shahar Gottlieb, Yoav Harris, and Doron Haviv. This is also where the most significant breakthrough was made – by an algorithm developed by Shahar Gottlieb – which led to the article's publication in Nature. Prof. Kaminer adds that the most interesting mathematical discovery made by the Ramanujan Machine's algorithms to date relates to a new algebraic structure concealed within a Catalan constant. The structure was discovered by high school student Yahel Manor, who participated in the project as part of the Alpha Program for science-oriented youth. Prof. Kaminer added that, "Industry colleagues Uri Mendlovic and Yaron Hadad also participated in the study, and contributed greatly to the mathematical and algorithmic concepts that form the foundation for the Ramanujan Machine. It is important to emphasize that the entire project was executed on a voluntary basis, received no funding, and participants joined the team out of pure scientific curiosity."

Prof. Ido Kaminer is the head of the Robert and Ruth Magid Electron Beam Quantum Dynamics Laboratory. He is a faculty member in the Andrew and Erna Viterbi Faculty of Electrical Engineering and the Solid State Institute. Kaminer is affiliated with the Helen Diller Quantum Centre and the Russell Berrie Nanotechnology Institute.

5.3 THE HARDY-RAMANUJAN AND ITS APPLICATION TO C++

The Hardy-Ramanujan Theorem states that the number of distinct prime factors of any natural number N will be approximately equal to the value of for most of the cases.

Question 1: What is a Prime number?

A prime number is a whole number greater than 1 that has only two factors: 1 and itself, meaning it cannot be divided evenly by any other numbers.

Common examples include **2, 3, 5, 7, 11, 13, 17, 19, 23, and 29**. The number 2 is the smallest and only even prime number.

Key Examples and Characteristics:

- **5 is Prime:** The only factors of 5 are 1 and 5 (1×5) =1×5=5).
- **13 is Prime:** It is not divisible by any integers other than 1 and 13.
- **97 is Prime:** It is a larger example, with only 1 and 97 as factors.
- **2 is Special: It is the only even prime number**; all other even numbers are divisible by 2, making them composite.
- **Non-Example:** 9 is not prime because it can be divided by 3 $(3 \times 3 = 9)$ 3×3=9), in addition to 1 and 9.
- The first few prime numbers are 2, 3, 5, 7, 11, 13, 17, 19, 23, 29, 31, 37, 41, 43, and 47.

Question 1: What is a Prime factor?

A prime factor is a prime number (like 2, 3, 5, 7) that divides another number exactly; for example, the prime factors of 12 are 2 and 3, because $12 = 2 \times 2 \times 3$=12, and 2 and 3 are prime numbers, while the prime factors of 20 are 2 and 5, since $20 = 2\times2\times5$. Prime factorization breaks a number down into these unique prime building blocks.

Examples :

- Number: 12:
  - Factors: 1, 2, 3, 4, 6, 12
  - Prime Factors: 2, 3 (because 2 and 3 are prime).
  - Prime Factorization: $2 \times 2 \times 3$.
- Number: 20:
  - Factors: 1, 2, 4, 5, 10, 20
  - Prime Factors: 2, 5 (4 and 10 are factors but not prime).
  - Prime Factorization: $2 \times 2 \times 5$.
- Number: 30:
  - Prime Factors: 2, 3, 5.
  - Prime Factorization: $2\times3\times5$

How to find them (Factor Tree Method Example for 36)

1. Start with the number (36).
2. Break it into any two factors, e.g., 2 and 18 (since $2\times18=36$).
3. Circle any prime numbers (2 is prime).
4. Continue factoring composite numbers (18) into smaller factors, e.g., 2 and 9 (since $2\times9=18$).
5. Circle the new prime (2) and factor the composite (9) into 3 and 3, circling both (3 is prime).
6. The prime factors are all the circled numbers: 2, 2, 3, 3.
7. Prime Factorization: $2 \times 2 \times 3 \times 3$.

- Prime Factors: Understand, Find & Master Math Basics Easily

For example, the factors of 12 are 1, 2, 3, 4, 6, and 12. Out of these, the prime numbers are 2 and 3. So, the prime factors of 12...

oVedantu

Related links

For example, let's consider N to be 1000.

The number of distinct prime factors of 15 are 2 and 5 i.e. 2 distinct prime factors.

The value of is equal to 1.932 which is approximately equal to 2.

The Hardy-Ramanujan theorem is proved in the above case.

Since the theorem states that the number of distinct prime factors will be approximately equal to for most of the cases. There are cases where this theorem does not imply.

For example, N=286

The number of distinct prime factors of N i.e. 286 are 2, 11, and 13. There are 3 distinct prime factors of 286.

According to Hardy-Ramanujan theorem, the number of distinct prime factors of N will be approximately equal to

$\log_e(\log_e(286))=1.732$, and check if they are approximately equal.

Implementation of Hardy-Ramanujan theorem in C++

To check the theorem for any natural number N, we will simply calculate all the distinct prime factors of N by creating a function to find the distinct prime factors of any number. We will count the number of distinct prime factors of N and then calculate the approximate number of distinct prime factors of N according to the Hardy-Ramanujan theorem i.e. equal to the value of

Syntax

```
int N;

int value=log(N);
```

To compare the approximate value of the number of distinct prime factors of any number according to Hardy-Ramanujan theorem with the actual number of distinct prime factors, we will create a function to count the number of distinct prime factors of any number.

We will check if the number N is divisible by 2, we will divide N by 2 until it gives 0 remainder and increase the count of distinct prime factors by 1. Now we will iterate in a for loop from i=3 to i<=sqrt(N) to check for every odd value of i as N will be odd at this time. If N is divisible by i, we will divide N by i until it gives remainder as 0 and update N everytime. Similarly, we will iterate and check for every value of i until it is less than or equal to the square root of N and update count by 1 for each iteration when i divides N.

Now for the case when N is greater than 2 and is a prime number, we will increase the count by 1 to get the exact number of distinct prime factors of N.

We will print the exact number of distinct prime factors calculated and approximate number of distinct prime factors of N according to Hardy-Ramanujan theorem to check the theorem for any natural number N.

The steps to implement the approach in C++:

- We will make a function to get the exact number of distinct prime factors of N which will be the user input as discussed above.
- We will print the exact number of distinct prime factors of N.

- Now we will calculate the approximate number of distinct prime factors of N by calculating the value of and store it in a variable.
- Print the approximate number of distinct prime factors of N according to the Hardy-Ramanujan theorem.

5.4 APPLICATION OF RAMANUJAN'S FINDING TO POLYMER CHEMISTRY AND SOME

Other Areas

The mathematical contributions of Srinivasa Ramanujan are applied in **polymer chemistry** to solve complex problems, particularly in understanding molecular structures, material properties, and synthesis. More than a century after his death, his work on partition theory, mock theta functions, and modular equations is utilized in this field.

Key applications in polymer chemistry and related material sciences include:

- **Polymer Structure and Analysis:** Ramanujan's formulas help in analysing the structural complexity of polymers, including the study of crystal lattices.
- **Nanotechnology & Composites:** His work is used in the study of polymer-based composites and nanocomposites, which are used to develop materials with improved mechanical strength, durability, and special conductivity properties.
- **Molecular Modeling:** His research is applied in understanding the behaviour of molecules, which is crucial for predicting the properties of new polymers.
- **Simulation and Modelling:** His techniques, such as the Ramanujan Master Theorem (used in Mellin transforms), are applied to analyse complex, data-driven systems.

5.5 RAMANUJAN'S FORMULAE FOR PI AND HIGH ENERGY PHYSICS

Most of us first hear about the irrational number π (pi)—rounded off as 3.14, with an infinite number of decimal digits—in school, where we learn about its use in the context of a circle. More recently, scientists have developed supercomputers that can estimate up to trillions of its digits.

Now, physicists at the Centre for High Energy Physics (CHEP), Indian Institute of Science (IISc), Bangalore, have found that pure mathematical formulas used to calculate the value of pi 100 years ago has connections to fundamental physics of today—showing up in theoretical models of **percolation**, **turbulence**, and certain aspects of **black holes.**

The research is published in the journal *Physical Review Letters*.

In 1914, just before he sailed from Madras to Cambridge, the famous Indian mathematician **Srinivasa Ramanujan** published a paper listing 17 mathematical formulas **to calculate pi**. They were highly efficient and helped compute pi faster than other methods at the time. Even with very few mathematical terms in them, the formulas still yielded many correct decimal digits of pi.

The formulas were so foundational that they form the basis for modern computational and mathematical techniques—even the ones used by supercomputers—to compute digits of pi.

"Scientists have computed pi up to 200 trillion digits using an algorithm called the Chudnovsky algorithm," says Aninda Sinha, Professor at CHEP and senior author of the new study. "These algorithms are actually based on Ramanujan's work."

5.5.1 Exploring the physics connection

The question that Sinha and Faizan Bhat, first author and former IISc Ph.D. student, asked was: Why should such astonishing formulas exist at all? In their work, they looked for a physics-based answer.

"We wanted to see whether the starting point of his formulas fit naturally into some physics," says Sinha. "In other words, is there a physical world where Ramanujan's mathematics appears on its own?"

They found that Ramanujan's formulas naturally come up within a broad class of theories called **conformal field theories**, specifically within logarithmic conformal field theories. Conformal field theories describe systems with scale invariance symmetry—essentially systems that look identical no matter how deep you zoom in, like fractals.

In a physical context, this can be seen at the critical point of water—a special temperature and pressure at which both liquid and vapor forms of water become indistinguishable from the other. At this point, water shows scale invariance symmetry and its properties can be described using conformal field theory.

Critical behaviour also comes up in **percolation** (how things spread through a medium), at the onset of **turbulence** in fluids, and certain descriptions of **black holes**—phenomena that can be explained by the more specific logarithmic conformal field theories.

5.5.2 Implications for modern physics

The researchers found that the mathematical structure underlying the starting point of Ramanujan's formulas also comes up in the mathematics underlying these logarithmic conformal field theories.

Using this connection, they could efficiently calculate certain quantities in these theories—ones that could potentially help them understand phenomena like turbulence or percolation better. This is similar to Ramanujan going from the starting point of his formulas and efficiently deriving pi.

"In any piece of beautiful mathematics, you almost always find that there is a physical system which actually mirrors the mathematics," says Bhat.

"Ramanujan's motivation might have been very mathematical, but without his knowledge, he was also studying black holes, turbulence, percolation, all sorts of things."

The study shows that Ramanujan's century-old formulas have a hitherto hidden application in making current high-energy physics calculations faster and more tractable. Even without this, however, Sinha and Bhat say they were just baffled by the beauty of Ramanujan's mathematics.

"We were simply fascinated by the way a genius working in early 20th century India, with almost no contact with modern physics, anticipated structures that are now central to our understanding of the universe," says Sinha.

5.5.3 Black Hole entropy

A separate concept in physics, entropy, explains why heat flows from a hot body to a cold body and not the other way around. The results of Ramanujan and Hardy on partitions and the former's subsequent work on what are called mock theta functions have come to play an important role in understanding the very quantum structure of spacetime – in particular the quantum entropy of a **type of Black Hole in string theory**, according to Prof. Dabholkar.

Stephen Hawking showed that when you take into account quantum effects, a Black Hole is not quite black, it is rather like a hot piece of metal that is slowly emitting Hawking radiation. Thus, one can associate thermodynamic quantities like temperature and entropy to a Black Hole.

"Mock modular forms are beginning to appear more and more in many areas of physics... Our work has also had unexpected applications in new topics in mathematics such as 'Umbral

Moonshine', which are quite unrelated to black holes," explains Prof. Dabholkar.

"It is a tribute to the remarkable originality and power of Ramanujan's genius that the ideas he created a century ago are now finding applications in such diverse contexts," he says.

5.6 RAMANUJAN'S LEGACY USED IN SIGNAL PROCESSING

There is no question about the fact that mathematical genius Srinivasa Ramanujan has left behind a rich legacy of problems for mathematicians to solve. In his short life of little over 32 years, he reached unimaginable heights. What is surprising is that his mathematics, done over a hundred years ago, finds applications today in areas other than pure mathematics, which were not even established during his time (22 December 1887 – 26 April 1920). One among these is **signal processing.**

Examples of signals that are processed digitally include obvious ones like speech and music and more research-oriented ones such as DNA and protein sequences. All these have certain patterns that repeat over and over again and are called periodic patterns. For example, a DNA molecule is made up of a 4 bases (Adenine Guanine, Thymine and Cytosine).

Sometimes, a sequence, say AGT, keeps repeating several times in a region of the DNA. In real life, more complex repeating patterns may need to be identified as they bear significance to health conditions. So, in signal processing, one thing we are interested in is extracting and identifying such periodic information.

Identifying and separating the periodic portion is much like using a sieve to separate particles of different sizes. A mathematical operation akin to a sieve is used to separate out the periodic regions in the signal. Some of the best-known methods to extract periodic components in signals involve Fourier analysis. Using Ramanujan Sums for this process is much less known.

"A Ramanujan Sum is a sequence like c(1), c(2), c(3) ... This sequence itself repeats periodically... It was thought, by a number of authors to be useful in identifying periodic components in signals, much the same as sines and cosines are used in Fourier analysis," says P.P. Vaidyanathan who has developed these ideas over the last decade. He is the Kiyo and Eiko Tomiyasu Professor of Electrical Engineering at the California Institute of Technology, U.S.

Prof. Vaidyanathan came across this work in a manner that illustrates how friendly connections play a role in the development of science. Several years ago, mathematicians H. Gopalakrishna Gadiyar and R. Padma, from VIT, Vellore, were studying the twin prime problem when they observed that some arithmetical function which captures the properties of the primes should have a **Ramanujan-Fourier Series.** They sent their paper to Bhaskar Ramamurthi, Director of IIT Madras, who in turn forwarded the paper to Prof. Vaidyanathan, a friend from his graduate days.

Intrigued by the Ramanujan Sum mentioned in the paper, Prof. Vaidyanathan delved deep into it and developed the concept of "Ramanujan subspaces." These ideas were further developed by his doctoral student Srikanth Tenneti who showed that using these gave a method that worked better than Fourier analysis when the region of periodicity is short.

5.7 APPLICATION OF RAMANUJAN'S MOCK THETA FUNCTIONS TO CANCER RESEARCH

Ramanujan's mock theta functions, originally derived as complex q-series, have found emerging applications in modelling the stochastic behaviour of oncogenes in genetic cancer. These functions, associated with harmonic Maass forms, help estimate the impact of viral and cellular oncogenes on tumour development.

5.7.1 Key Connections and Applications:

(i) **Oncogene Modelling:** Research utilizes mock theta functions, such as

$X_0(q)c{-}erb$ and = $X_0(q)$ $c{-}sis$ to analyse the growth of specific oncogenes (e.g., $c{-}erb$, $c{-}sis$, $c{-}fms$) and their role in cancer.

(ii) Tumour Development: The functions assist in evaluating the conditions

($[F_0(q)]v_0$+$[F_0(q)]c_0$ > $u(-q)$) under which viral and cellular oncogenes cause, for example, bronchial carcinoma.

Bibliography

Agarwal. R. P. "Resonance of Ramanujan's Mathematics", New Age International Publishers, Volume 1, 1996.

Alladi, Krishnaswami. "Ramanujan's Place in the World of Mathematics: Essays Providing a Comparative Study", Springer India, 2013.

Andrews, George E. "Srinivasa Ramanujan The Lost Note Book And Other Unpublished Papers" 1988 New Delhi Narosa Publishing House, New Delhi, 1988.

Berndt, Bruce C. "Ramanujan's Notebooks, Part-1", From G. H. Hardy, Ramanujan, Twelve Lectures on Subjects Suggested by His Life and Work. Cambridge University Press, 1940. Springer-Verlag New York, 1985.

Faizan Bhat et al, "Ramanujan's $1/\pi$ Series and Conformal Field Theories", Physical Review Letters (2025). DOI: 10.1103/c38g-fd2v

Fitouhi, Ahmed; Kamel Brahim; and Neji Bettaibi. "On some q-versions of Ramanujan master theorem", The Ramanujan Journal 50(3), November 2019.DOI:10.1007/s11139-019-00141-4

Hardy, G. H. "Ramanujan: Twelve Lectures on Subjects suggested by his Life and Work", Cambridge University Press, 1940.

http://tiff.net/festivals/festival15/galapresentations/the-man-who-knew-infinity. "The Man Who Knew Infinity". Toronto International Film Festival. Directed by Matthew Brown, Created by United Kingdom | 114 minutes | World Premiere Artist: Dev Patel and Academy Award winner Jeremy Irons star in this inspirational biopic about Srinivasa Ramanujan, the early-20th century Indian mathematician whose groundbreaking theories revolutionized the field.

Kanigel, Robert. "The Man Who Knew Infinity: A life of the Genius Ramanujan', Washington Square Press, New York, 1991.

Ramanujan, Srinivasa. "The Lost Note Book And Other Unpublished Papers" Narosa Publishing House, New Delhi, 1988.

Ranganathan, S. R. "Ramanujan: The Man & the Mathematician, Ess Ess Publications, Bangalore. 2009

Srinivas, Sydney. "Srinivasa Ramanujan(English)", Prism Books.

Venkatesh, Arundhati and Priya Kuriyan' "Ramanujan: From Zero to Infinity", Penguin Random House India Pvt Ltd. 2023.

Vengallur, Krishnachandran. "Ramanujan in Computing Technology", March 2021

DOI:10.48550/arXiv.2103.09654

Waldschmidt, Michel. "Srinivasa Ramanujan: His life and his work", Khon Kaen University November 2016. Institut de Math´ematiques de Jussieu & Paris VI. http://www.imj-prg.fr/~michel.waldschmidt/

www.ingramcontent.com/pod-product-compliance
Lightning Source LLC
LaVergne TN
LVHW052115210926
848482LV00025B/631
* 9 7 9 8 9 0 3 6 2 5 6 7 3 *